Ben Stacy Jerrik (Ed.)

Niederbipp

Ben Stacy Jerrik (Ed.)

Niederbipp

Oberaargau, Communes of Switzerland, Cantons of Switzerland

Part Press

Publisher:
Part Press is a trademark of
International Book Market Service Ltd., 17 Rue Meldrum, Beau Bassin, 1713-01 Mauritius
Email: info@bookmarketservice.com
Website: www.bookmarketservice.com

Published in 2013

Printed in: U.S.A., U.K., Germany. This book was not produced in Mauritius.

ISBN: 978-620-1-61263-1

Contents

Articles

References

Hunters_Hollow,_Kentucky

Hunters Hollow is a 6th-class city in Bullitt County, Kentucky, in the United States. The population was 372 at the time of the 2000 U.S. census.

Geography

Hunters Hollow is located at (38.078251, -85.693975).[] According to the United States Census Bureau, the city has a total area of 0.1 square miles (km^2), all land.

Demographics

As of the census[] of 2000, there were 372 people, 117 households, and 106 families residing in the city. The population density was 5,778.3 people per square mile (2,393.8/km²). There were 119 housing units at an average density of 1,848.4 per square mile (765.8/km²). The racial makeup of the city was 97.58% White, 0.54% Asian, 0.54% from other races, and 1.34% from two or more races. Hispanic or Latino of any race were 0.27% of the population.

There were 117 households out of which 52.1% had children under the age of 18 living with them, 76.1% were married couples living together, 12.0% had a female householder with no husband present, and 9.4% were non-families. 6.8% of all households were made up of individuals and 1.7% had someone living alone who was 65 years of age or older. The average household size was 3.18 and the average family size was 3.33.

In the city the population was spread out with 31.7% under the age of 18, 7.8% from 18 to 24, 38.4% from 25 to 44, 20.7% from 45 to 64, and 1.3% who were 65 years of age or older. The median age was 31 years. For every 100 females there were 91.8 males. For every 100 females age 18 and over, there were 96.9 males.

The median income for a household in the city was $55,000, and the median income for a family was $55,750. Males had a median income of $38,375 versus $20,809 for females. The per capita income for the city was $17,665. About 1.8% of families and 3.0% of the population were below the poverty line, including 5.6% of those under age 18 and none of those age 65 or over.

References

Bullitt_County,_Kentucky

Bullitt County, founded in December 1796,[1] is a county in the U.S. state of Kentucky[] located in the far western Bluegrass region known as the Knobs, with an area of 300 square miles (km^2). The population was 74,319 in the 2010 Census. Just south of the city of Louisville, part of the Louisville metropolitan area commonly known as Kentuckiana, the county is home to several commuter towns including the cities of Hillview, Lebanon Junction, Mount Washington and Shepherdsville - the county seat. The western fifth of the county (**unknown operator: u','** acres (**unknown operator: u'strong'** km^2)) is reserved for military training as part of the United States Army post of Fort Knox.[2]

History

The first inhabitants of the land that would become Bullitt County were the Paleo-Indians who entered North America approximately 11,500 to 10,000 years BP. These people, whose ancestors can be traced back to Central Asia, were nomadic, hunters and gatherers whose remains have been discovered near the area's mineral springs or salt licks where big game such as the mammoth, bison and ground sloth once gathered. Native Americans, including the Shawnee people who probably considered this region part of their homeland and certainly valued it as a hunting ground, were their descendants. [3] [4]

European colonization of the Americas led to competing claims to the lands west of the Appalachians and east of the Mississippi river with France ultimately ceding control to Britain at the end of the French and Indian War in 1763. Veterans of the war were promised land and in 1773 Captain Thomas Bullitt was sent on a surveying expedition of the area around the Falls of the Ohio by the governor of Virginia. During his explorations of the land that would become Bullitt County, he came upon an unusually large salt lick subsequently named after him: Bullitt's Lick. [5]

Bullitt's Lick became an important saltwork to the region - supplying salt by pack train and flatboat as far off as Illinois. In fact, the Bullitt's Lick saltwork was Kentucky's first industry and in production until around 1830, when the steamboat and importation brought access to less expensive sources of salt.

The first settlement of the area and the first station on the Wilderness Road between Harrodsburg and the Falls of the Ohio was a fort called Brashear's Station or the Salt River Garrison, built in 1779 at the mouth of Floyd's Fork. Shepherdsville, named after Adam Shepherd - a prosperous business man who purchased the land near the Falls of Salt River in 1793, is the oldest town and became the county seat.

In December of 1796[1] the county of Bullitt - named after Thomas Bullitt's nephew and Kentucky's first Lieutenant Governor Alexander Scott Bullitt - was created from land given by Jefferson and Nelson counties through an act approved on December 13, 1796 by the Kentucky General Assembly. In 1811, the northwestern area of the county expanded to include land given by Jefferson County. In 1824, an eastern area of the county was given to help form Spencer County.[]

Geography

Topography

According to the 2000 census, the county has a total area of 300.20 square miles (km^2), of which 299.08 square miles (km^2) (or 99.63%) is land and 1.12 square miles (km^2) (or 0.37%) is water.[6]

Adjacent counties

- Jefferson County - (north)
- Spencer County - (east)
- Nelson County - (southeast)
- Hardin County - (southwest)

Cities

Although large-scale residential development has not made its way south of Shepherdsville, the growth is apparent in and around that town and in Mount Washington, as well as points north along I-65 towards Hillview. Both Shepherdsville and Mount Washington have stretched their boundaries such that they are nearly touching each other. The 10-mile (km) stretch of Kentucky 44 that connects the two towns has homes through nearly the entire expanse.

- Brooks
- Brownington
- Clermont
- Fox Chase
- Hebron Estates
- Hillview
- Hunters Hollow
- Lebanon Junction
- Mount Washington
- Pioneer Village
- Shepherdsville
- Solitude

Demographics

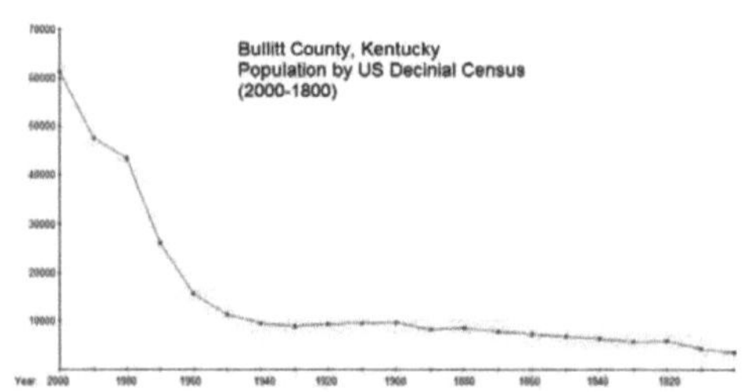

Graph of Bullitt County population by decade

Historical populations		
Census	Pop.	%±
1800	3542	—
1810	4311	21.7%
1820	5831	35.3%
1830	5652	−3.1%
1840	6334	12.1%
1850	6774	6.9%
1860	7289	7.6%
1870	7781	6.7%
1880	8521	9.5%
1890	8291	−2.7%
1900	9602	15.8%
1910	9487	−1.2%
1920	9328	−1.7%
1930	8868	−4.9%
1940	9511	7.3%
1950	11349	19.3%
1960	15726	38.6%
1970	26090	65.9%
1980	43346	66.1%
1990	47567	9.7%
2000	61236	28.7%
2010	74319	21.4%
Est. 2012	75896	2.1%
U.S. Decennial Census[7] 2012 Estimate[8]		

As of the census[] of 2000, there were 61,236 people, 22,171 households, and 17,736 families residing in the county. The population density was 205 per square mile (/km^2). There were 23,160 housing units at an average density of 77 per square mile (/km^2). The racial makeup of the county was 98.07% White, 0.38% Black or African American, 0.34% Native American, 0.27% Asian, 0.01% Pacific Islander, 0.16% from other races, and 0.77% from two or more races. 0.63% of the population were Hispanics or Latinos of any race.

There were 22,171 households out of which 39.00% had children under the age of 18 living with them, 65.40% were married couples living together, 10.40% had a female householder with no husband present, and 20.00% were non-families. 16.40% of all households were made up of individuals and 5.40% had someone living alone who was 65 years of age or older. The average household size was 2.75 and the average family size was 3.07.

In the county the population was spread out with 27.20% under the age of 18, 8.60% from 18 to 24, 32.70% from 25 to 44, 23.70% from 45 to 64, and 7.80% who were 65 years of age or older. The median age was 34 years. For every 100 females there were 98.90 males. For every 100 females age 18 and over, there were 97.20 males.

The median income for a household in the county was $50,058 (2005), and the median income for a family was $49,481. Males had a median income of $35,851 versus $24,098 for females. The per capita income for the county was $18,339. About 6.20% of families and 7.90% of the population were below the poverty line, including 11.40% of those under age 18 and 7.60% of those age 65 or over.

Economy

A UPS 757 Freighter aircraft

Bullitt County, which is bisected by I-65, the main north-south transportation corridor, has grown into a thriving distribution hub, and several of its major business parks are approximately 16 miles (km) from Louisville International Airport and UPS' global air-freight hub Worldport. More than 6000000 square feet (m^2) of distribution, warehousing and other complexes have been built and absorbed in Bullitt County since 2000, and additional construction is ongoing.

In October 2011, Zappos, which already is Bullitt County's largest employer, with more than 2,300 full-time workers, was reported to have plans to add 5,000 more employees in the next three years. The company has also outgrown the more than 1 million square feet of processing space it operates in Bullitt County and plans to build a 1.2 million-square-foot facility next year to keep pace with its rapid growth. [9]

Bullitt County residents have easy access to major job centers such as Elizabethtown, Fort Knox, and Louisville.

Law and government

Public safety

There are several police agencies in Bullitt County. The primary law enforcement agency in the county is the Bullitt County Sheriff.[10] The current Sheriff of Bullitt County is Dave Greenwell. The Sheriff's Office is an elected position and is staffed by 34 Deputy Sheriff's, and four office staff. There are also 9 Deputies and Court Certified Security Officers (CCSOs) who provide courthouse security as required by Kentucky law. The Sheriff's office provides patrol, crime prevention, criminal investigation; and all other police related functions in the county. The primary areas for the Sheriff's patrol division are the unincorporated areas in Bullitt County they also provide backup for the various city police agencies. Since 1989 the Sheriff's Office has been contracted by the County Fiscal Court to provide the County Police force to the citizens of Bullitt County after the County Police Department was folded. The County Sheriff's Office has deputies assigned to the Drug Task Force, Arson Task Force, Accident Reconstruction Team, Rapid Response Team, Technical Operations, ATV Squad, Boat Patrol, Hostage Negotiations, and an Incident Command Team. Most of the services are provided while on duty or for free by Special Deputies who serve their county for free, even paying for their own equipment, uniforms, and sidearm.

Fire protection is provided by both Kentucky Chapter 75 and 95 Districts and Departments. Zoneton Fire Protection District, Mount Washington Fire Protection District, and the Shepherdsville Division of Fire are currently staffed with a 24-hour professional firefighting staff. Nichols, Southeast Bullitt, and Lebanon Junction are all volunteer-based.

- Zoneton Fire District
- Mt. Washington Fire District
- Nichols Fire District

- Southeast Bullitt Fire Department
- Shepherdsville Fire Department
- Lebanon Junction Fire Department

The Bullitt County Emergency Medical Service (BCEMS) provides all emergency medical care and transport in Bullitt County. BCEMS is a division of Bullitt County Fiscal Court and authorized and funded by the county through taxes.

Education

Primary and secondary schools

Bullitt County is served by Bullitt County Public Schools.

There are six middle schools:

- Bernheim Middle School
- Bullitt Lick Middle School
- Eastside Middle School
- Hebron Middle School
- Mount Washington Middle School
- Zoneton Middle School

There are four high schools:

- Bullitt Central High School, located in Shepherdsville
- Bullitt East High School, located in Mount Washington
- North Bullitt High School, located in Hebron Estates (served by the Shepherdsville post office)
- Riverview High School

See also

- National Register of Historic Places listings in Bullitt County, Kentucky

References

[1] http://www.kyenc.org/entry/b/BULLI05.html
[2] Kempf, Gary. *The Land Before Fort Knox.* Arcadia Publishing, 2004, p. 11.
[10] Bullitt County Sheriff (http://www.bullittcountysheriff.com)

External links

- Bernheim Arboretum and Research Forest (http://www.Bernheim.org/)
- Bullitt County government and commerce website (http://www.bullittcounty.ky.gov/)
- Bullitt County History Museum (http://www.bullittcountyhistory.org/)
- Bullitt County Attractions (http://www.travelbullitt.org/)

United_States

United States of America

Motto:

"In God we trust" (official)[1] [2] [3]

"E pluribus unum" (Latin) (traditional)
"Out of many, one"

Anthem: "The Star-Spangled Banner"

Capital		Washington D.C.
Largest city		New York City
Official languages		None at federal level[a]
National language		English[b]
Demonym		American
Government		Federal presidential constitutional republic
-	President	Barack Obama (D)
-	Vice President	Joe Biden (D)
-	Speaker of the House	John Boehner (R)
-	Chief Justice	John Roberts
Legislature		Congress
-	Upper house	Senate
-	Lower house	House of Representatives
Independence from the Kingdom of Great Britain		
-	Declared	July 4, 1776

-	Recognized	September 3, 1783
-	Current constitution	June 21, 1788
	Area	
-	Total	9826675 km^2[] [c] (3rd/4th) 3794101 sq mi
-	Water (%)	6.76
	Population	
-	2013 estimate	316438000[4] (3rd)
-	Density	34.2/km^2 (179th) 88.6/sq mi
GDP (PPP)		2012 estimate
-	Total	$15.685 trillion[] (1st)
-	Per capita	$49,922[] (6th)
GDP (nominal)		2012 estimate
-	Total	$15.685 trillion[] (1st)
-	Per capita	$49,922[] (12th)
Gini (2011)		47.7[5] high · 39th (2009)
HDI (2013)		▲ 0.937[] very high · 3rd
	Currency	[[]] ($) (USD)
	Time zone	(UTC−5 to −10)
-	Summer (DST)	(UTC−4 to −10[e])
	Drives on the	right[g]
	Calling code	+1
	ISO 3166 code	US
	Internet TLD	.us.gov.mil.edu
a.	English is the official language of at least 28 states; some sources give higher figures, based on differing definitions of "official".[] English and Hawaiian are both official languages in the state of Hawaii.	
b.	English is the *de facto* language of American government and the sole language spoken at home by 80 percent of Americans aged five and older. Spanish is the second most commonly spoken language.	
c.	Whether the United States or China is larger is disputed. The figure given is from the U.S. Central Intelligence Agency's *The World Factbook*. Other sources give smaller figures. All authoritative calculations of the country's size include only the 50 states and the District of Columbia, not the territories.	
d.	The population estimate is of people whose usual residence is within the 50 states and the District of Columbia, regardless of nationality. It does not include those living in the territories (over 4 million people, mostly in Puerto Rico).	
e.	See Time in the United States for details about laws governing time zones in the United States.	
f.	Does not include insular areas and United States Minor Outlying Islands, which have their own ISO 3166 codes.	
g.	Except US Virgin Islands.	

The **United States of America** (**USA** or **U.S.A.**), commonly referred to as the **United States** (**US** or **U.S.**) and **America**, is a federal republic[6] [7] consisting of 50 states and a federal district. The 48 contiguous states and the federal district of Washington, D.C. are in central North America between Canada and Mexico. The state of Alaska is the northwestern part of North America, west of Canada and east of Russia which is across the Bering Strait in Asia, and the state of Hawaii is an archipelago in the mid-North Pacific. The country also has five populated and nine unpopulated territories in the Pacific and the Caribbean.

At 3.79 million square miles (9.83 million km^2) in total and with around 315 million people, the United States is the third largest country by land area and population. It is one of the world's most ethnically diverse and multicultural nations, the product of large-scale immigration from many countries.[8] The geography and climate of the U.S. is also extremely diverse, with deserts, plains, forests, and mountains that are home to a wide variety of wildlife.

Paleo-indians migrated from Asia to what is now the United States mainland around 12,000 years ago. European colonization began around 1600 and came mostly from England. The United States emerged from 13 British colonies located along the Atlantic seaboard. Disputes between Great Britain and the American colonies led to the American Revolution. On July 4, 1776, delegates from the 13 colonies unanimously issued the Declaration of Independence, which established the United States of America. The American Revolutionary War, which ended with the recognition of independence of the United States from the Kingdom of Great Britain, was the first successful war of independence against a European colonial empire.[9] [10] The current Constitution was adopted on September 17, 1787; 27 Amendments have since been added to the Constitution. The first 10 amendments, collectively named the Bill of Rights, were ratified in 1791 and guarantee many fundamental civil rights and freedoms.

Driven by the doctrine of manifest destiny, the United States embarked on a vigorous expansion across North America throughout the 19th century.[] This involved displacing native tribes, acquiring new territories, and gradually admitting new states.[] The American Civil War ended legalized slavery in the United States.[11] By the end of the 19th century, the United States extended into the Pacific Ocean,[] and its economy was the world's largest.[12] The Spanish–American War and World War I confirmed the country's status as a global military power. The United States emerged from World War II as a global superpower, the first country with nuclear weapons, and a permanent member of the United Nations Security Council. The end of the Cold War and the dissolution of the Soviet Union left the United States as the sole superpower.

The United States is a developed country and has the world's largest national economy, with an estimated 2013 GDP of $16.2 trillion –22% of global GDP at purchasing-power parity, as of 2011.[] [] [13] The per capita GDP of the U.S. was the world's sixth-highest as of 2010.[] A 2013 study ranked the US as having the fourth most unequal income distribution among OECD nations,[14] though it has the highest mean and second highest median household income in the OECD, as well as the highest average employee compensation.[] [] The economy is fueled by an abundance of natural resources, a well-developed infrastructure,[15] and high productivity;[16] and while its economy is considered post-industrial it continues to be one of the world's largest manufacturers.[17] The country accounts for 39% of global military spending,[18] being the foremost economic and military power, a prominent political and cultural force in the world, and a leader in scientific research and technological innovation.[19] [20]

Etymology

In 1507, German cartographer Martin Waldseemüller produced a world map on which he named the lands of the Western Hemisphere "America" after Italian explorer and cartographer Amerigo Vespucci.[21]

The first documentary evidence of the phrase "United States of America" is from a letter dated January 2, 1776, written by Stephen Moylan, Esq., George Washingon's aide-de-camp and Muster-Master General of the Continental Army. Addressed to Col. Joseph Reed, Moylan expressed his wish to carry the "full and ample powers of the United States of America" to Spain to assist in the revolutionary war effort.[22]

The first publicly published evidence of the phrase "United States of America" was in an anonymously written essay in *The Virginia Gazette* newspaper in Williamsburg, Virginia on April 6, 1776.[23] [24] In June 1776, Thomas

Jefferson included the phrase "UNITED STATES OF AMERICA" in all capitalized letters in the headline of his "original Rough draught" of the Declaration of Independence.[25] [26] In the final Fourth of July version of the Declaration, the pertinent section of the title was changed to read, "The unanimous Declaration of the thirteen united States of America".[27]

In 1777 the Articles of Confederation announced, "The Stile of this Confederacy shall be 'The United States of America'".[28]

The short form "United States" is also standard. Other common forms include the "U.S.", the "USA", and "America". Colloquial names include the "U.S. of A." and, internationally, the "States". "Columbia", a name popular in poetry and songs of the late 1700s,[29] derives its origin from Christopher Columbus; it appears in the name "District of Columbia".

The standard way to refer to a citizen of the United States is as an "American". "United States", "American" and "U.S." are used to refer to the country adjectivally ("American values", "U.S. forces"). "American" is rarely used in English to refer to subjects not connected with the United States.[30]

The phrase "United States" was originally treated as plural, a description of a collection of independent states—e.g., "the United States are"—including in the Thirteenth Amendment to the United States Constitution, ratified in 1865. It became common to treat it as singular, a single unit—e.g., "the United States is"—after the end of the Civil War. The singular form is now standard; the plural form is retained in the idiom "these United States".[] The difference has been described as more significant than one of usage, but reflecting the difference between a collection of states and a unit.[31]

In non-English languages, the name is frequently translated as the translation of either the "United States" or "United States of America", and colloquially as "America". In addition, an initialism is sometimes used.[32]

History

Native American and European contact

1766 engraving of a meeting between Native Americans and Europeans

People from Asia migrated to the North American continent approximately 12,000 or more years ago.[33] Some, such as the pre-Columbian Mississippian culture, developed advanced agriculture, grand architecture, and state-level societies. After European explorers and traders made the first contacts, it is estimated that their population declined due to various reasons, including diseases such as smallpox and measles to which indigenous Americans had no natural immunities,[34] [35] intermarriage,[36] and violence.[37] [38] [39]

In the early days of colonization many settlers were subject to shortages of food, disease and attacks from native Indians. Indians were also often at war with neighboring tribes and would often enslave their defeated enemy, a practice that was also soon used by various colonists who captured Indians in battle. During the various colonial wars, many colonists were also captured by Indians as slaves and taken north to Canada and sold to the French.[40]

At the same time however many natives and settlers got along and came to depend on each other, especially settlers during the winter months. Natives also came to depend on settlers for guns, ammunition, powder and other modern devices. Because many tribes were frequently at war with one another it became imperative to establish and secure good relationships with at least one group of colonists. As colonists began to spread out into the interior their contact with native Indians increased, sometimes resulting in good relations, oftentimes resulting in conflict. In the process "Native American influenced colonist, and colonist influenced Native American".[41]

Natives taught many settlers where, when and how to hunt and fish in the vast frontier that lay before them whose elements were generally unknown to the Europeans. In order to survive settlers often depended on native Indians who taught them how to adopt to the Indian's "hunting culture" and learned the use of animal skins as camouflage, decoys along with various whistles and calls used to attract prey. European ministries and others felt it was important to "civilize" the Indians and urged them to concentrate on farming and ranching and not depend primarily on hunting and gathering. At the same time Indians offered the benefit of their experience in growing corn, an unknown crop in Europe, and in the use of dead fish and other methods as fertilizer. It was not long before many Indians began to grow new crops and raise livestock and poultry in their communities and made use of the various living utilities settlers had to offer.[42] [43]

Initially the Puritan and Wampanoag were peaceful, however the King Philip's War began following cultural and religious differences between the colonists and the Wampanoag;[44] [45] by the war's end, the European colonists had defeated the Native Americans and were able to expand and control New England.[46] In Carolina, Native Americans were captured and sold into slavery to both New England and the West Indies. In 1676, the Virginia colony legally sanctioned the enslavement of Native Americans.[47] Conversely, the Five Civilized Tribes were involved in the institution of African slavery as planters.[48]

Settlements

Further information: European colonization of the Americas and 13 colonies

The signing of the Mayflower Compact, 1620.

After Columbus' discovery of the New World in 1492 other explorers followed.[49] The first Spanish explorers landed in "La Florida" in 1513. Conquistadors explored much of the continent's interior and Spain later set up some settlements in parts of Florida and the American southwest that were eventually merged into the United States.[50] There were also some French attempts to colonize the east coast, and later more successful settlements along the Mississippi River. Many early European colonies failed due to starvation, disease, harsh weather, Indian attacks, or warfare with European rivals. The fate of the "lost" English colony of Roanoke in the 1580s is an enduring mystery.

James I on April 10, 1606 chartered The Virginia Company with the purpose of establishing English settlements on the eastern coast of North America. The Virginia Colony was planted in 1607 with Jamestown and the Pilgrims' Plymouth Colony in 1620. Both colonies suffered initial hardships and great loss of life, but eventually stabilized and became the first successful English settlements in America. Both also saw efficiency greatly improve when personal property replaced the early communal operations.[51] The continent's first elected legislative assembly, Virginia's House of Burgesses created in 1619, and the Mayflower Compact, signed by the Pilgrims before disembarking, established precedents for the pattern of representative self-government and constitutionalism that would develop throughout the American colonies.[52] [53] Tens of thousands of Puritans later settled New England.

Other New England colonies were established. Much of the territory between them and Virginia was controlled by the Dutch until England seized it in the late 17th century during the Anglo-Dutch Wars, leading to the creation of the Middle Colonies.[54] Trade with and Christian evangelism to local Indians were established in the colonies' early days, though relations would alternate from friendly to tense, and were characterized by periodic bouts of warfare, often with some tribes allying themselves with the English against common foes. Incidents like the massacre of 1622, the Pequot War, and King Philip's War caused great destruction and threatened the existence of entire colonies, but resulted in reprisals that ultimately saw the power of enemy tribes reduced or broken, facilitating the expansion of English settlements.[55] [56]

Most settlers in every colony were small farmers, but other industries developed. Tobacco was popular in Europe and became a major early cash crop. Furs, fishing, lumber, rum, rice, indigo, construction, wheat, ranching, and eventually shipbuilding contributed to economic growth. By the late colonial period Americans were producing one-seventh of the world's iron supply.[57] Cities eventually dotted the coast to support local economies and serve as trade hubs. English colonists were supplemented by waves of Scotch-Irish and other groups. As coastal land grew more expensive people pushed west into the hills and backwoods, seeking to carve an existence out of virgin wilderness.[58]

Settlers were a diverse mix of adventurers, profit seekers, people wanting religious freedom, and those who simply saw an opportunity for a better life.[59] Many came as indentured servants, either convicts or people who otherwise couldn't afford passage voluntarily signing contracts, and were set free after completing their specified term of service. Two-thirds of all Virginia settlers between 1630 and 1680 arrived indentured.[60]

Penn's Treaty between the Quaker settlers and Native Americans was never violated during William Penn's lifetime.

The first African slaves were brought to the Americas by Spanish conquistadors in the 1500s shortly after Columbus' voyages. Most slaves were shipped to sugar colonies in the Caribbean and to Brazil, where life expectancy was about seven years.[61] Life expectancy was much higher in North America because of less disease and better food and treatment, so the numbers of slaves grew rapidly into the millions by excesses of births over deaths.[62] [63]

Colonial society was largely divided over the religious and moral implications of slavery and many colonies passed acts for and against the practice.[64] [65] By the turn of the 18th century, African slaves were becoming the primary source of bonded labor in many regions.[66] Some colonists participated in the lucrative, slave oriented "Golden Triangle", involving planters, merchants of various types, shippers, and the African tribal chiefs who provided them with slaves.[47] [67]

With the 1729 division of the Carolinas and the 1732 colonization of Georgia, the 13 colonies that would become the United States of America were established.[] All had local governments with elections open to most free men, with a growing devotion to the ancient rights of Englishmen and a sense of self-government stimulating support for republicanism.[] With extremely high birth rates, low death rates, and steady settlement, the colonial population grew rapidly. Relatively small Indian populations were eclipsed.[68] The Christian revivalist movement of the 1730s and 1740s known as the Great Awakening fueled interest in both religion and religious liberty.

In the French and Indian War, British forces seized Canada from the French, but the francophone population remained politically isolated from the southern colonies. Excluding the Native Americans, who were being conquered and displaced, those 13 colonies had a population of over 2.1 million in 1770, about one-third that of Britain. Despite continuing new arrivals, the rate of natural increase was such that by the 1770s only a small minority of Americans had been born overseas.[69] The colonies' distance from Britain had allowed the development of self-government, but their success motivated monarchs to periodically seek to reassert Royal authority.

Independence and expansion

Further information: American Revolutionary War, Declaration of Independence, and American Revolution

The American Revolution was the first successful colonial war of independence against a European power. Americans had developed a democratic system of local government and an ideology of "republicanism" that held government rested on the will of the people (not the king), which strongly opposed corruption and demanded civic virtue. They demanded their rights as Englishmen and rejected British efforts to impose taxes without the approval of colonial legislatures. The British insisted and the conflict escalated to full-scale war in 1775, the American Revolutionary War.[] On June 14, 1775, the Continental Congress, convening in Philadelphia, established a Continental Army under the command of George Washington.[] Proclaiming that "all men are created equal" and endowed with "certain unalienable Rights", the Congress adopted the Declaration of Independence, drafted largely by Thomas Jefferson, on July 4, 1776. That date is now celebrated annually as America's Independence Day. In

1777, the Articles of Confederation established a weak government that operated until 1789.[]

The *Declaration of Independence*: the Committee of Five presenting their draft to the Second Continental Congress in 1776.

U.S. Territories were acquired, organized with citizens, republican self-government, then admitted states on equal footing.

After a naval victory followed by the British defeat at Yorktown by American forces assisted by the French,[70] the United States was independent. In the peace treaty of 1783 Britain recognized American sovereignty over most territory east of the Mississippi River. Nationalists calling for a much stronger federal government with powers of taxation led the constitutional convention in 1787. After intense debate in state conventions the United States Constitution was ratified in 1788. The first Senate, House of Representatives, and president—George Washington—took office in 1789. The Bill of Rights, forbidding federal restriction of personal freedoms and guaranteeing a range of legal protections, was adopted in 1791.[71]

Attitudes toward slavery were shifting; nearly all states officially outlawed the international slave trade before the federal government criminalized it in 1808.[] Slavery had become more pronounced in the south than the north because the land there was better suited for large scale cash crop cultivation than the rocky ground and cooler climate of New England.[72] [73] All the Northern states abolished slavery between 1780 and 1804, leaving the slave states of the South as defenders of the "peculiar institution". With cotton a highly profitable plantation crop after 1820, slave interests in the Southern states maintained that slavery was a positive good for everyone, including the slaves.[] The Second Great Awakening, beginning about 1800, converted millions to evangelical Protestantism. In the North it energized multiple social reform movements, including abolitionism.[]

Americans' eagerness to expand westward prompted a long series of Indian Wars.[] The Louisiana Purchase of French-claimed territory under President Thomas Jefferson in 1803 almost doubled the nation's size.[74] The War of 1812, declared against Britain over various grievances and fought to a draw, strengthened U.S. nationalism.[] A series of U.S. military incursions into Florida led Spain to cede it and other Gulf Coast territory in 1819.[]

President Andrew Jackson took office in 1829, and began a set of reforms which led to the era of Jacksonian democracy, which is considered to have lasted from 1830 to 1850. This included many reforms, such as wider male suffrage, and various adjustments to the power of the Federal government. This also led to the rise of the Second Party System, which refers to the dominant parties which existed from 1828 to 1854.

The Trail of Tears in the 1830s exemplified the Indian removal policy that moved Indians to their own reservations, sometimes by force, with small annual government subsidies. The United States annexed the Republic of Texas in 1845, amid a period when the concept of Manifest Destiny was becoming popular.[] The 1846 Oregon Treaty with Britain led to U.S. control of the present-day American Northwest.[] The U.S. victory in the Mexican-American War resulted in the 1848 cession of California and much of the present-day American Southwest.[]

The California Gold Rush of 1848–49 further spurred western migration.[] New railways made relocation easier for settlers and increased conflicts with Native Americans.[] Over a half-century, up to 40 million American bison, or buffalo, were slaughtered for skins and meat and to ease the railways' spread.[] The loss of the buffalo, a primary resource for the plains Indians, was an existential blow to many native cultures.[] In 1869, President Ulysses S. Grant's Peace policy reversed the previous costly policy of "wars of extermination" in order to civilize and give Indians eventual United State citizenship having incorporated Indians as wards of the state, led by a philanthropic Board of Indian Commissioners.[75]

Civil War and Reconstruction Era

Further information: American Civil War and Reconstruction Era

Starting in the 1780s inherent divisions between the North and the South in American society over slavery ultimately led to the American Civil War.[76] Initially, the Founders of the nation had been able to keep the Union solvent by compromises worked out at the Constitutional Convention and to remain a single nation.[76]

Battle of Gettysburg, Pennsylvania. The American Civil WarCivil War cemented the Union and spurred the steel industry and First Transcontinental Railroadintercontinental railroad construction.

During the years leading up to the American Civil War tensions between slave and free states mounted with arguments about the relationship between the state and federal governments, as well as violent conflicts over the spread of slavery into new states.[] Abraham Lincoln, candidate of the largely antislavery Republican Party, was elected president in 1860.[] Before he took office, seven slave states declared their secession—which the federal government maintained was illegal—and formed the Confederate States of America.[]

With the Confederate attack upon Fort Sumter, the Civil War began and four more slave states joined the Confederacy.[] Lincoln's Emancipation Proclamation in 1863 declared slaves in the Confederacy to be free, though not those in Union slave states. Following the Union victory in 1865, three amendments to the U.S. Constitution ensured freedom for the nearly four million African Americans who had been slaves,[77] made them citizens, and gave them voting rights. The war and its resolution led to a substantial increase in federal power.[78] The war remains the deadliest conflict in American history, resulting in the deaths of 620,000 soldiers.[79]

The assassination of Abraham Lincoln radicalized Republican Reconstruction policies aimed at reintegrating and rebuilding the Southern states while ensuring the rights of the newly freed slaves.[] President Ulysses S. Grant implemented the Department of Justice and used the U.S. Military to enforce suffrage and civil rights for African Americans in the South destroying the Ku Klux Klan in 1871 under the Force Acts.[80] The resolution of the disputed 1876 presidential election by the Compromise of 1877 ended Reconstruction; Jim Crow laws soon disenfranchised many African Americans.[]

Industrialization

In the North, urbanization and an unprecedented influx of immigrants from Southern and Eastern Europe hastened the country's industrialization. The wave of immigration, lasting until 1924, provided labor and transformed American culture.[] United States immigration policies were Eurocentric, which barred Asians from naturalization, and restricted their immigration beginning with the Chinese Exclusion Act in 1882.[81] National infrastructure development spurred economic growth. The end of the Civil War spurred greater settlement and development of the American Old West. This was due to a variety of social and technological developments, including the completion of the First Transcontinental Telegraph in 1861 and the First Transcontinental Railroad in 1869.

Ellis Island, New York City. East Coast of the United StatesEast Coast immigrants worked in factories, railroads, and mines, and created demand for industrialized agriculture.

The 1867 Alaska Purchase from Russia completed the country's mainland expansion. The Wounded Knee Massacre in 1890 was the last major armed conflict of the Indian Wars. In 1893, the indigenous monarchy of the Pacific Kingdom of Hawaii was overthrown in a coup led by American residents; the United States annexed the archipelago in 1898. Victory in the Spanish–American War the same year demonstrated that the United States was a world power and led to the annexation of Puerto Rico, Guam, and the Philippines.[82] The Philippines gained independence a half-century later; Puerto Rico and Guam remain U.S. territories.

The emergence of many prominent industrialists at the end of the 19th century gave rise to the Gilded Age, a period of growing affluence and power among the business class. The hardships imposed upon the working classes during this period led to the rise of anarchist and socialist movements in the US.[83] In 1914 alone, 35,000 workers died in industrial accidents and 700,000 were injured.[84] This period eventually ended with the beginning of the Progressive Era, a period of great reforms in many societal areas, including regulatory protection for the public, greater antitrust measures, and attention to living conditions for the working classes. President Theodore Roosevelt was one leading proponent of progressive reforms.

World War I, Great Depression, and World War II

Further information: World War I, Great Depression, and World War II

At the outbreak of World War I in 1914, the United States remained neutral. Most Americans sympathized with the British and French, although many opposed intervention.[85] In 1917, the United States joined the Allies, and the American Expeditionary Forces helped to turn the tide against the Central Powers. President Woodrow Wilson took a leading diplomatic role at the Paris Peace Conference of 1919 which helped to shape the post-war world. Wilson advocated strongly for the U.S. to join the League of Nations. However, the Senate refused to approve this, and did not ratify the Treaty of Versailles, which established the League of Nations.[86]

The Dust Bowl brought agricultural depression, impacted industrial markets, and led to large relocation out of the Great Plains.

WWII invasion of Europe required war industry, accelerating migration to big cities and large scale manufacturing

The country pursued a policy of unilateralism, verging on isolationism.[86] In 1920, the women's rights movement, led by Carrie Chapman Catt, won passage of a constitutional amendment granting women's suffrage.[] The prosperity of the Roaring Twenties ended with the Wall Street Crash of 1929 that triggered the Great Depression.

After his election as president in 1932, Franklin D. Roosevelt responded with the New Deal, a range of policies increasing government intervention in the economy, including the establishment of the Social Security system.[87] The Dust Bowl of the mid-1930s impoverished many farming communities and spurred a new wave of western migration.

The United States, effectively neutral during World War II's early stages after Nazi Germany's invasion of Poland in September 1939, began supplying material to the Allies in March 1941 through the Lend-Lease program. On December 7, 1941, the Empire of Japan launched a surprise attack on Pearl Harbor, prompting the United States to join the Allies against the Axis powers as well as the internment of Japanese Americans by the thousands.[88] Participation in the war spurred capital investment and industrial capacity, and the production figures after the Americans started to unfold the awesome productive capacity of their economy became the stuff of legend.[89] Though the nation lost more than 400,000 soldiers,[90] among the major combatants, the United States was the only nation to become richer because of the war.[91]

Allied conferences at Bretton Woods and Yalta outlined a new system of international organizations that placed the United States and Soviet Union at the center of world affairs. As victory was won in Europe, a 1945 international conference held in San Francisco produced the United Nations Charter, which became active after the war.[92] The United States, having developed the first nuclear weapons, used them on the Japanese cities of Hiroshima and Nagasaki in August. Japan surrendered on September 2, ending the war.[93]

Cold War and Civil Rights era

Civil Rights leaders, including Ralph Abernathy and Martin Luther King, Jr., lead one of the Selma to Montgomery marches

A vehicle enters West Berlin through the Western Allies' Checkpoint Charlie after the fall of the Berlin Wall, marking the beginning of the end of the Cold War

The United States and the Soviet Union jockeyed for power after World War II during the Cold War, dominating the military affairs of Europe through NATO and the Warsaw Pact, respectively. While they engaged in proxy wars and developed powerful nuclear arsenals, the two countries avoided direct military conflict. The U.S. often opposed

Third World left-wing movements that it viewed as Soviet-sponsored. American troops fought Communist Chinese and North Korean forces in the Korean War of 1950–53. The House Un-American Activities Committee pursued a series of investigations into suspected leftist subversion, while Senator Joseph McCarthy became the figurehead of anticommunist sentiment.[94]

The 1961 Soviet launch of the first manned spaceflight prompted President John F. Kennedy's call for the United States to be first to land "a man on the moon", achieved in 1969.[95] Kennedy also faced a tense nuclear showdown with Soviet forces in Cuba.[96] Meanwhile, the United States experienced sustained economic expansion. Amidst the presence of various white nationalist groups, particularly the Ku Klux Klan, a growing civil rights movement used nonviolence to confront segregation and discrimination. This was symbolized and led by black Americans such as Rosa Parks and Martin Luther King, Jr. On the other hand, some black nationalist groups such as the Black Panther Party and Malcolm X had a more militant scope.

Following Kennedy's assassination in 1963, the Civil Rights Act of 1964, Voting Rights Act of 1965, and Immigration and Nationality Act of 1965 were passed under President Lyndon B. Johnson.[97] He also signed into law the Medicare and Medicaid programs.[98] Johnson also expanded a proxy war in Southeast Asia into the ultimately unsuccessful Vietnam War. A widespread countercultural movement grew, fueled by opposition to the war, black nationalism, and the sexual revolution. Betty Friedan, Gloria Steinem, and others led a new wave of feminism that sought political, social, and economic equality for women.

As president, Richard Nixon ended U.S. involvement in the Vietnam War, improved relations with China, and oversaw the beginning of a period of generally eased relations with the Soviet Union. As a result of the Watergate scandal, in 1974 Nixon became the first U.S. president to resign, to avoid being impeached on charges including obstruction of justice and abuse of power. The Jimmy Carter administration of the late 1970s was marked by stagflation and the Iran hostage crisis. The election of Ronald Reagan as president in 1980 heralded a rightward shift in American politics,[99] [100] [101] [102] reflected in major changes in taxation and spending priorities.[103] His second term in office brought both the Iran–Contra scandal and significant diplomatic progress with the Soviet Union.[104] The subsequent Soviet collapse ended the Cold War.[105] [106] [107] [108] [109]

Contemporary era

Under President George H. W. Bush, the United States took a lead role in the UN–sanctioned Gulf War.[110] The longest economic expansion in modern U.S. history—from March 1991 to March 2001—encompassed the Bill Clinton administration and the dot-com bubble.[111]

September 11, 2001 attacks on the World Trade Center in New York City.

On September 11, 2001, al-Qaeda terrorists under the leadership of Osama bin Laden struck the World Trade Center in New York City and The Pentagon near Washington, D.C., killing nearly 3,000 people.[112] In response, the George W. Bush administration launched the global War on Terror, invading Afghanistan and removing the Taliban government and al-Qaeda training camps.[113] Bin Laden was killed by a Navy SEAL raid in Pakistan in 2011;[114] Taliban insurgents continue to fight a guerrilla war.[115] In 2003, the United States and several allied forces invaded Iraq to engineer regime change there; the war lasted eight years.[116] [117] [118] In 2005, Hurricane Katrina caused severe destruction along much of the Gulf Coast, devastating New Orleans.[119]

In 2008, amid a global economic recession, the first African-American president, Barack Obama, was elected.[120]

Geography, climate, and environment

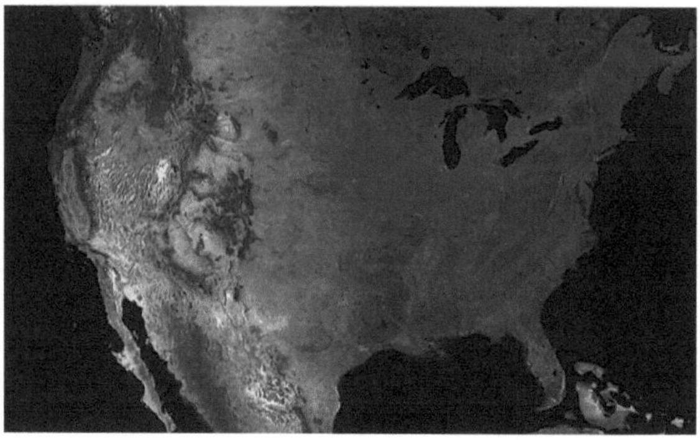
A composite satellite image of the contiguous United States

The land area of the contiguous United States is 2959064 square miles (km^2). Alaska, separated from the contiguous United States by Canada, is the largest state at 663268 square miles (km^2). Hawaii, occupying an archipelago in the central Pacific, southwest of North America, is 10931 square miles (km^2) in area.[121]

The United States is the world's third or fourth largest nation by total area (land and water), ranking behind Russia and Canada and just above or below China. The ranking varies depending on how two territories disputed by China and India are counted and how the total size of the United States is measured: calculations range from 3676486 square miles (km^2)[122] to 3717813 square miles (km^2)[123] to 3794101 square miles (km^2).[] Measured by only land area, the United States is third in size behind Russia and China, just ahead of Canada.[124]

The coastal plain of the Atlantic seaboard gives way further inland to deciduous forests and the rolling hills of the Piedmont. The Appalachian Mountains divide the eastern seaboard from the Great Lakes and the grasslands of the Midwest. The Mississippi–Missouri River, the world's fourth longest river system, runs mainly north–south through the heart of the country. The flat, fertile prairie of the Great Plains stretches to the west, interrupted by a highland region in the southeast.

A Bald Eagle, the national bird of the United States since 1782.

The Rocky Mountains, at the western edge of the Great Plains, extend north to south across the country, reaching altitudes higher than 14,000 feet (4,300 m) in Colorado. Farther west are the rocky Great Basin and deserts such as the Chihuahua and Mojave. The Sierra Nevada and Cascade mountain ranges run close to the Pacific coast, both ranges reaching altitudes higher than 14000 feet (m).

The lowest and highest points in the continental United States are in the state of California, and only about 80 miles (km) apart. At 20,320 feet (6,194 m), Alaska's Mount McKinley is the tallest peak in the country and in North America. Active volcanoes are common throughout Alaska's Alexander and Aleutian Islands, and Hawaii consists of volcanic islands. The supervolcano underlying Yellowstone National Park in the Rockies is the continent's largest volcanic feature.[125]

The United States, with its large size and geographic variety, includes most climate types. To the east of the 100th meridian, the climate ranges from humid continental in the north to humid subtropical in the south. The southern tip of Florida is tropical, as is Hawaii. The Great Plains west of the 100th meridian are semi-arid. Much of the Western mountains are alpine. The climate is arid in the Great Basin, desert in the Southwest, Mediterranean in coastal California, and oceanic in coastal Oregon and Washington and southern Alaska. Most of Alaska is subarctic or polar. Extreme weather is not uncommon—the states bordering the Gulf of Mexico are prone to hurricanes, and most of the world's tornadoes occur within the country, mainly in the Midwest's Tornado Alley.[126]

The U.S. ecology is considered "megadiverse": about 17,000 species of vascular plants occur in the contiguous United States and Alaska, and over 1,800 species of flowering plants are found in Hawaii, few of which occur on the mainland.[127] The United States is home to more than 400 mammal, 750 bird, and 500 reptile and amphibian species.[128] About 91,000 insect species have been described.[129]

There are 58 national parks and hundreds of other federally managed parks, forests, and wilderness areas.[130] Altogether, the government owns 28.8% of the country's land area.[] Most of this is protected, though some is leased for oil and gas drilling, mining, logging, or cattle ranching; 2.4% is used for military purposes.[] [131] []

Environmental issues

Environmental issues have been high on the national agenda since 1970. The controversies include debates on oil and nuclear energy, dealing with air and water pollution, the economic costs of protecting wildlife, and international responses to global warming.[132] [133]

Many federal and state agencies are involved. The most prominent is the Environmental Protection Agency (EPA), created by presidential order in 1970.[134] The idea of wilderness has shaped the management of public lands since 1964, with the Wilderness Act.[135] The Endangered Species Act of 1973 protects threatened and endangered species and their habitats, which are monitored by the United States Fish and Wildlife Service.

Demographics

Population

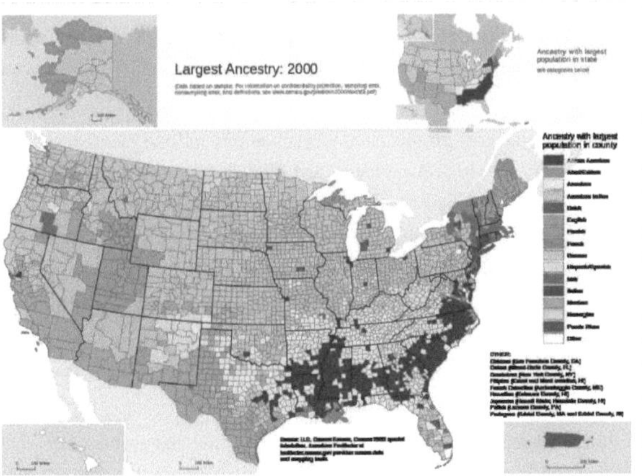

Largest ancestry groups by county, 2000

Race/Ethnicity	
(as given by the 2010 Census)[]	
By race:	
White	72.4%
Black American	12.6%
Asian	4.8%
American Indian and Alaska Native	0.9%
Native Hawaiian and Pacific Islander	0.2%
Other	6.2%
Multiracial (2 or more)	2.9%
By ethnicity:[136]	
Hispanic/Latino (of any race)	16.3%
Non-Hispanic/Latino (of any race)	83.7%

The U.S. Census Bureau estimates the country's population now to be 316438000,[4] including an approximate 11.2 million illegal aliens.[137] The U.S. population almost quadrupled during the 20th century, from about 76 million in 1900.[138] The third most populous nation in the world, after China and India, the United States is the only major industrialized nation in which large population increases are projected.[]

With a birth rate of 13 per 1,000, 35% below the world average, its population growth rate is positive at 0.9%, significantly higher than those of many developed nations.[139] In fiscal year 2012, over one million immigrants (most of whom entered through family reunification) were granted legal residence.[140] Mexico has been the leading source of new residents for over two decades; since 1998, China, India, and the Philippines have been in the top four sending countries every year.[141] [142] 9 million Americans identify as lesbian, gay, bisexual or transgender, making up four percent of the population.[143] A 2010 survey found that seven percent of men and eight percent of women identified as gay, lesbian or bisexual.[144]

The United States has a very diverse population—31 ancestry groups have more than one million members.[] White Americans are the largest racial group; German Americans, Irish Americans, and English Americans constitute three of the country's four largest ancestry groups.[] Black Americans are the nation's largest racial minority and third largest ancestry group.[] Asian Americans are the country's second largest racial minority; the three largest Asian American ethnic groups are Chinese Americans, Filipino Americans, and Indian Americans.[]

In 2010, the U.S. population included an estimated 5.2 million people with some American Indian or Alaska Native ancestry (2.9 million exclusively of such ancestry) and 1.2 million with some native Hawaiian or Pacific island ancestry (0.5 million exclusively).[] The census counted more than 19 million people of "Some Other Race" who were "unable to identify with any" of its five official race categories in 2010.[]

The population growth of Hispanic and Latino Americans (the terms are officially interchangeable) is a major demographic trend. The 50.5 million Americans of Hispanic descent[] are identified as sharing a distinct "ethnicity" by the Census Bureau; 64% of Hispanic Americans are of Mexican descent.[] Between 2000 and 2010, the country's Hispanic population increased 43% while the non-Hispanic population rose just 4.9%.[] Much of this growth is from immigration; as of 2007, 12.6% of the U.S. population was foreign-born, with 54% of that figure born in Latin America.[145]

Fertility is also a factor; as of 2010 the average Hispanic (of any race) woman gave birth to 2.35 children in her lifetime, compared to 1.97 for non-Hispanic black women and 1.79 for non-Hispanic white women (both below the replacement rate of 2.1).[] Minorities (as defined by the Census Bureau as all those beside non-Hispanic, non-multiracial whites) constituted 36.3% of the population in 2010,[146] and over 50% of children under age one,[147] and are projected to constitute the majority by 2042.[148] This contradicts the report by the National Vital Statistics Reports, based on the U.S. census data, which concludes that, 54% (2,162,406 out of 3,999,386 in 2010) of births were non-Hispanic white.[]

About 82% of Americans live in urban areas (including suburbs);[] about half of those reside in cities with populations over 50,000.[149] In 2008, 273 incorporated places had populations over 100,000, nine cities had more than one million residents, and four global cities had over two million (New York City, Los Angeles, Chicago, and Houston).[]

There are 52 metropolitan areas with populations greater than one million.[] Of the 50 fastest-growing metro areas, 47 are in the West or South.[150] The metro areas of Dallas, Houston, Atlanta, and Phoenix all grew by more than a million people between 2000 and 2008.[]

Leading population centers (see complete list)				
Rank	**Core city (cities)**	**Metro area population**	**Metropolitan Statistical Area**	**Region**[151]
1	**New York City**	19,015,900	New York–New Jersey–Connecticut–Pennsylvania, NY–NJ–CT–PA MSA	Northeast
2	**Los Angeles**	12,944,801	Los Angeles–Long Beach–Santa Ana, CA MSA	West
3	**Chicago**	9,504,753	Chicago–Joliet–Naperville, IL–IN–WI MSA	Midwest
4	**Dallas–Fort Worth**	6,526,548	Dallas–Fort Worth–Arlington, TX MSA	South
5	**Houston**	6,086,538	Houston–The Woodlands-Sugar Land MSA	South
6	**Philadelphia**	5,992,414	Philadelphia–Camden–Wilmington, PA–NJ–DE–MD MSA	Northeast
7	**Washington, D.C.**	5,703,948	Washington, DC–VA–MD–WV MSA	South
8	**Miami**	5,670,125	Miami–Fort Lauderdale–Pompano Beach, FL MSA	South
9	**Atlanta**	5,359,205	Atlanta–Sandy Springs–Marietta, GA MSA	South
10	**Boston**	4,591,112	Boston–Cambridge–Quincy, MA–NH MSA	Northeast
11	**San Francisco**	4,391,037	San Francisco–Oakland–Fremont, CA MSA	West
12	**San Bernardino-Riverside**	4,304,997	San Bernandino–Riverside–Ontario, CA MSA	West
13	**Detroit**	4,285,832	Detroit–Warren–Livonia, MI MSA	Midwest
14	**Phoenix**	4,263,236	Phoenix–Mesa–Glendale, AZ MSA	West
15	**Seattle**	3,500,026	Seattle–Tacoma–Bellevue, WA MSA	West
16	**Minneapolis–St. Paul**	3,318,486	Minneapolis–St. Paul–Bloomington, MN–WI MSA	Midwest
17	**San Diego**	3,140,069	San Diego–Carlsbad–San Marcos, CA MSA	West
18	**Tampa–St. Petersburg**	2,824,724	Tampa–St. Petersburg–Clearwater, FL MSA	South
19	**St. Louis**	2,817,355	St. Louis–St. Charles–Farmington, MO–IL MSA	Midwest
20	**Baltimore**	2,729,110	Baltimore–Towson, MD MSA	South
based upon 2011 population estimates from the U.S. Census Bureau[152]				

New York City

Los Angeles

Chicago

Language

Languages (2010)[153]	
English (*only*)	229.7 million
Spanish, incl. Creole	37.0 million
Chinese	2.8 million
French, incl. Creole	2.1 million
Tagalog	1.6 million
Vietnamese	1.4 million
Italian	1.1 million
Korean	1.1 million
German	1.1 million

English (American English) is the de facto national language. Although there is no official language at the federal level, some laws—such as U.S. naturalization requirements—standardize English. In 2010, about 230 million, or 80% of the population aged five years and older, spoke only English at home. Spanish, spoken by 12% of the population at home, is the second most common language and the most widely taught second language.[153] [154] Some Americans advocate making English the country's official language, as it is in at least 28 states.[]

Both Hawaiian and English are official languages in Hawaii, by state law.[155] While neither has an official language, New Mexico has laws providing for the use of both English and Spanish, as Louisiana does for English and French.[156] Other states, such as California, mandate the publication of Spanish versions of certain government documents including court forms.[157] Many jurisdictions with large numbers of non-English speakers produce government materials, especially voting information, in the most commonly spoken languages in those jurisdictions.

Several insular territories grant official recognition to their native languages, along with English: Samoan and Chamorro are recognized by American Samoa and Guam, respectively; Carolinian and Chamorro are recognized by the Northern Mariana Islands; Spanish is an official language of Puerto Rico and is more widely spoken than English there.

Religion

The First Amendment of the U.S. Constitution guarantees the free exercise of religion and forbids Congress from passing laws respecting its establishment. In a 2002 study, 59% of Americans said that religion played a "very important role in their lives", a far higher figure than that of any other wealthy nation.[158] According to a 2007 survey, 78.4% of adults identified themselves as Christian,[] down from 86.4% in 1990.[]

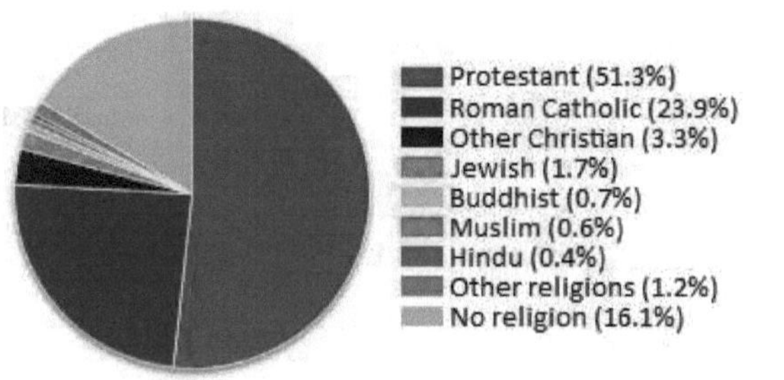

A pie chart of religious groups in the U.S. (2007)

Protestant denominations accounted for 51.3%, while Roman Catholicism, at 23.9%, was the largest individual denomination.[] The total reporting non-Christian religions in 2007 was 4.7%, up from 3.3% in 1990.[] Other religions include Judaism (1.7%), Buddhism (0.7%), Islam (0.6%), Hinduism (0.4%), and Unitarian Universalism

(0.3%).[] The survey also reported that 16.1% of Americans described themselves as agnostic, atheist, or simply having no religion, up from 8.2% in 1990.[] [] []

There are also Baha'i, Sikh, Jain, Shinto, Confucian, Taoist, Druid, Native American, Wiccan, humanist and deist communities.[159] Doubt about the existence of a god or gods is growing rapidly among Americans under 30.[160] Polls show that overall American confidence in organized religion is declining,[161] and that younger Americans in particular are becoming increasingly irreligious.[162]

Family structure

In 2007, 58% of Americans age 18 and over were married, 6% were widowed, 10% were divorced, and 25% had never been married.[163] Women now work mostly outside the home and receive a majority of bachelor's degrees.[164]

The U.S. teenage pregnancy rate, 79.8 per 1,000 women, is the highest among OECD nations.[165] Between 2007 and 2010, the highest teenage birth rate was in Mississippi, and the lowest in New Hampshire.[166] While the abortion rate is falling, the abortion ratio of 241 per 1,000 live births and abortion rate of 15 per 1,000 women aged 15–44 remain higher than those of most Western nations.[167]

Government and politics

U.S. Capitol seats the Congress: the Senate, left; the House, right

The White House used by the U.S. President

Supreme Court and offices for nine Justices

The United States is the world's oldest surviving federation. It is a constitutional republic and representative democracy, "in which majority rule is tempered by minority rights protected by law".[168] The government is regulated by a system of checks and balances defined by the U.S. Constitution, which serves as the country's supreme legal document.[169]

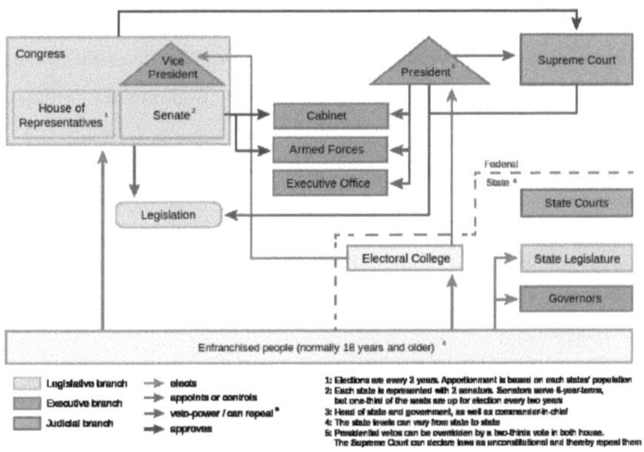

Political system of the United States

In the American federalist system, citizens are usually subject to three levels of government: federal, state, and local. The local government's duties are commonly split between county and municipal governments. In almost all cases, executive and legislative officials are elected by a plurality vote of citizens by district. There is no proportional representation at the federal level, and it is very rare at lower levels.

The federal government is composed of three branches:

- Legislative: The bicameral Congress, made up of the Senate and the House of Representatives, makes federal law, declares war, approves treaties, has the power of the purse,[170] and has the power of impeachment, by which it can remove sitting members of the government.[171]

- Executive: The president is the commander-in-chief of the military, can veto legislative bills before they become law (subject to Congressional override), and appoints the members of the Cabinet (subject to Senate approval) and other officers, who administer and enforce federal laws and policies.[172]
- Judicial: The Supreme Court and lower federal courts, whose judges are appointed by the president with Senate approval, interpret laws and overturn those they find unconstitutional.

The House of Representatives has 435 voting members, each representing a congressional district for a two-year term. House seats are apportioned among the states by population every tenth year. As of the 2010 census, seven states have the minimum of one representative, while California, the most populous state, has 53.[173]

The Senate has 100 members with each state having two senators, elected at-large to six-year terms; one third of Senate seats are up for election every other year. The president serves a four-year term and may be elected to the office no more than twice. The president is not elected by direct vote, but by an indirect electoral college system in which the determining votes are apportioned to the states and the District of Columbia.[174] The Supreme Court, led by the Chief Justice of the United States, has nine members, who serve for life.[175]

The state governments are structured in roughly similar fashion; Nebraska uniquely has a unicameral legislature.[176] The governor (chief executive) of each state is directly elected. Some state judges and cabinet officers are appointed by the governors of the respective states, while others are elected by popular vote.

The original text of the Constitution establishes the structure and responsibilities of the federal government and its relationship with the individual states. Article One protects the right to the "great writ" of habeas corpus, The Constitution has been amended 27 times;[177] the first 10 amendments, which make up the Bill of Rights, and the Fourteenth Amendment form the central basis of Americans' individual rights. All laws and governmental procedures are subject to judicial review and any law ruled in violation of the Constitution is voided. The principle of judicial review, not explicitly mentioned in the Constitution, was declared by the Supreme Court in *Marbury v. Madison* (1803),[178] a decision handed down by Chief Justice John Marshall.[179]

Political divisions

Further information: Territorial evolution of the United States and United States territorial acquisitions

The United States is a federal union of 50 states. The original 13 states were the successors of the 13 colonies that rebelled against British rule. Early in the country's history, three new states were organized on territory separated from the claims of the existing states: Kentucky from Virginia; Tennessee from North Carolina; and Maine from Massachusetts. Most of the other states have been carved from territories obtained through war or purchase by the U.S. government. One set of exceptions includes Vermont, Texas, and Hawaii: each was an independent republic before joining the union. During the American Civil War, West Virginia broke away from Virginia. The most recent state—Hawaii—achieved statehood on August 21, 1959.[180] The states do not have the right to unilaterally secede from the union.

The states compose the vast bulk of the U.S. land mass; the two other areas considered integral parts of the country are the District of Columbia, the federal district where the capital, Washington, is located; and Palmyra Atoll, an uninhabited but incorporated territory in the Pacific Ocean. The United States also possesses five major overseas territories: Puerto Rico and the United States Virgin Islands in the Caribbean; and American Samoa, Guam, and the Northern Mariana Islands in the Pacific.[181] Those born in the major territories are birthright U.S. citizens except Samoans. Samoans born in American Samoa are born U.S. nationals, and may become naturalized citizens.[182] American citizens residing in the territories have fundamental constitutional protections and elective self-government, with a territorial Member of Congress, but they do not vote for president as states. Territories have personal and business tax regimes different from that of states.[183]

The United States also observes tribal sovereignty of the Native Nations. Though reservations are within state borders, the reservation is a sovereign. While the United States recognizes this sovereignty, other countries may not.[184]

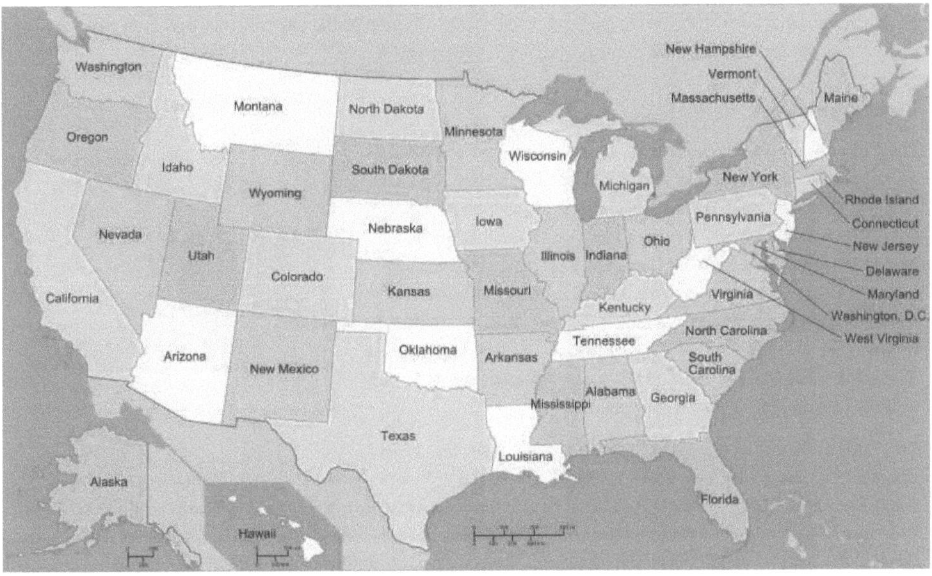

Parties and elections

(from left to right) House Majority Leader Eric Cantor, House Minority Leader Nancy Pelosi, House Speaker John Boehner, President Barack Obama, Senate Majority Leader Harry Reid, Senate Minority Leader Mitch McConnell at the White House in 2011

The United States has operated under a two-party system for most of its history.[] For elective offices at most levels, state-administered primary elections choose the major party nominees for subsequent general elections. Since the general election of 1856, the major parties have been the Democratic Party, founded in 1824, and the Republican Party, founded in 1854. Since the Civil War, only one third-party presidential candidate—former president Theodore Roosevelt, running as a Progressive in 1912—has won as much as 20% of the popular vote. The third-largest political party is the Libertarian Party.

Within American political culture, the Republican Party is considered center-right or conservative and the Democratic Party is considered center-left or liberal.[185] The states of the Northeast and West Coast and some of the Great Lakes states, known as "blue states", are relatively liberal. The "red states" of the South and parts of the Great Plains and Rocky Mountains are relatively conservative.

The winner of the 2008 presidential election and the 2012 presidential election, Democrat Barack Obama, is the 44th U.S. president.

In the 113th United States Congress, the House of Representatives is controlled by the Republican Party, while the Democratic Party has control of the Senate. The Senate currently consists of 52 Democrats, two independents who caucus with the Democrats, and 46 Republicans; the House consists of 234 Republicans and 201 Democrats.[186] There are 30 Republican and 19 Democratic state governors, as well as one independent.[187]

Since the founding of the United States until 2000s, the country's governance has been primarily dominated by White Anglo-Saxon Protestants (WASPs). However, the situation has changed recently and of the top 17 positions (four national candidates of the two major party in the 2012 U.S. presidential election, four leaders in 112th United States Congress, and nine Supreme Court Justices) there is only one WASP.[188] [189] [190]

Foreign relations

The United States has established foreign relations. It is a permanent member of the United Nations Security Council, and New York City hosts the United Nations Headquarters. It is a member of the G8,[191] G20, and Organisation for Economic Co-operation and Development. Almost all countries have embassies in Washington, D.C., and many have consulates around the country. Likewise, nearly all nations host American diplomatic missions. However, Cuba, Iran, North Korea, Bhutan, and the Republic of China (Taiwan) do not have formal diplomatic relations with the United States (although the U.S. still supplies Taiwan with military equipment).

British Foreign Secretary William Hague and U.S. Secretary of State Hillary Clinton, May 2010

The United States has a "special relationship" with the United Kingdom[192] and strong ties with Canada,[193] Australia,[194] New Zealand,[195] the Philippines,[196] Japan,[197] South Korea,[198] Israel,[199] and several European countries such as Russia, France and Germany. It works closely with Russia and fellow NATO members on military and security issues and with its neighbors through the Organization of American States and free trade agreements such as the trilateral North American Free Trade Agreement with Canada and Mexico. In 2008, the United States spent a net $25.4 billion on official development assistance, the most in the world. As a share of America's large gross national income (GNI), however, the U.S. contribution of 0.18% ranked last among 22 donor states. By contrast, private overseas giving by Americans is relatively generous.[200]

Government finance

Taxes are levied in the United States at the federal, state and local government level. These include taxes on income, payroll, property, sales, imports, estates and gifts, as well as various fees. In 2010 taxes collected by federal, state and municipal governments amounted to 24.8% of GDP.[201] During FY2012, the federal government collected approximately $2.45 trillion in tax revenue, up $147 billion or 6% versus FY2011 revenues of $2.30 trillion. Primary receipt categories included individual income taxes ($1,132B or 47%), Social Security/Social Insurance taxes ($845B or 35%), and corporate taxes ($242B or 10%).[]

US taxation is generally progressive, especially the federal income taxes, and is among the most progressive in the developed world.[202] [] [203] [204] [205] In 2009 the top 10% of earners, with 36% of the nation's income, paid 78.2% of the federal personal income tax burden, while the bottom 40% had a negative liability.[] Payroll taxes are less progressive, as the entitlement programs they ostensibly fund have historically not been viewed as welfare transfers.[206] [207] The top 10% paid 51.8% of total federal taxes in 2009, and the top 1%, with 13.4% of pre-tax national income, paid 22.3% of federal taxes.[] In 2013 the Tax Policy Center projected total federal effective tax rates of 35.5% for the top 1%, 27.2% for the top quintile, 13.8% for the middle quintile, and −2.7% for the bottom quintile.[208] [209] State and local taxes vary widely, but are generally less progressive than federal taxes as they rely heavily on broadly borne regressive sales and property taxes that yield less volatile revenue streams, though their consideration doesn't eliminate the progressive nature of overall taxation.[210] [211]

During FY 2012, the federal government spent $3.54 trillion on a budget or cash basis, down $60 billion or 1.7% vs. FY 2011 spending of $3.60 trillion. Major categories of FY 2012 spending included: Medicare & Medicaid ($802B or 23% of spending), Social Security ($768B or 22%), Defense Department ($670B or 19%), non-defense discretionary ($615B or 17%), other mandatory ($461B or 13%) and interest ($223B or 6%).[]

Public debt

In March 2013, US federal government debt held by the public was approximately $11.888 trillion, or about 75% of US GDP. Intra-governmental holdings stood at $4.861 trillion, giving a combined total debt of $16.749 trillion.[] [] By 2012 total federal debt had surpassed 100% of US GDP.[212] The US has a credit rating of AA+ from Standard & Poor's, AAA from Fitch, and Aaa from Moody's.[213]

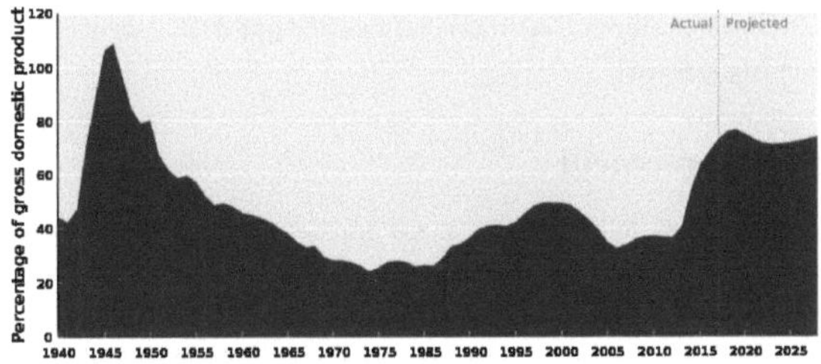

U.S. federal debt held by the public as a percentage of GDP, from 1940 to 2012.

Historically, the US public debt as a share of GDP increased during wars and recessions, and subsequently declined. For example, debt held by the public as a share of GDP peaked just after World War II (113% of GDP in 1945), but then fell over the following 30 years. In recent decades, large budget deficits and the resulting increases in debt have led to concern about the long-term sustainability of the federal government's fiscal policies.[214] However, these concerns are not universally shared.[215]

Military

The president holds the title of commander-in-chief of the nation's armed forces and appoints its leaders, the secretary of defense and the Joint Chiefs of Staff. The United States Department of Defense administers the armed forces, including the Army, Navy, Marine Corps, and Air Force. The Coast Guard is run by the Department of Homeland Security in peacetime and the Department of the Navy in time of war. In 2008, the armed forces had 1.4 million personnel on active duty. The Reserves and National Guard brought the total number of troops to 2.3 million. The Department of Defense also employed about 700,000 civilians, not including contractors.[216]

Military service is voluntary, though conscription may occur in wartime through the Selective Service System.[217] American forces can be rapidly deployed by the Air Force's large fleet of transport aircraft, the Navy's 11 active aircraft carriers, and Marine Expeditionary Units at sea with the Navy's Atlantic and Pacific fleets. The military operates 865 bases and facilities abroad,[218] and maintains deployments greater than 100 active duty personnel in 25 foreign countries.[219] The extent of this global military presence has prompted some scholars to describe the United States as maintaining an "empire of bases".[220] The United States is in many ways a new kind of entity in the geopolitical bestiary: a destabilizing hegemon, a revolutionary empire.[]

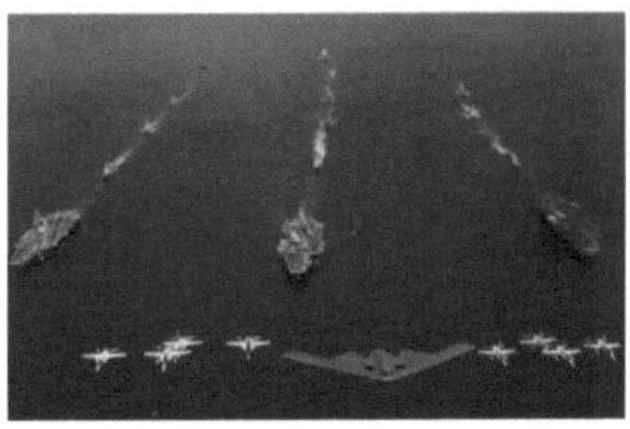
The carrier strike groups of the *Kitty Hawk*, *Ronald Reagan*, and *Abraham Lincoln* with aircraft from the Marine Corps, Navy, and Air Force.

Total U.S. military spending in 2011, more than $700 billion, was 41% of global military spending and equal to the next 14 largest national military expenditures combined. At 4.7% of GDP, the rate was the second-highest among the top 15 military spenders, after Saudi Arabia.[221] U.S. defense spending as a percentage of GDP ranks 23rd globally as of 2012 according to the CIA.[222] Defense's share of U.S. spending has generally declined in recent decades, from

Cold War peaks of 14.2% of GDP in 1953 and 69.5% of federal outlays in 1954 to 4.7% of GDP and 18.8% of federal outlays in 2011.[223]

The proposed base Department of Defense budget for 2012, $553 billion, is a 4.2% increase over 2011; an additional $118 billion is proposed for the military campaigns in Iraq and Afghanistan.[224] The last American troops serving in Iraq departed in December 2011;[225] 4,484 servicemen were killed during the Iraq War.[226] Approximately 90,000 U.S. troops were serving in Afghanistan as of April 2012;[] as of May 21, 2013, 2,039 had been killed during the War in Afghanistan.[227]

Law enforcement

Law enforcement in the U.S. is maintained primarily by local police departments. The New York City Police Department (NYPD) is the largest in the country.[228]

Law enforcement in the United States is primarily the responsibility of local police and sheriff's departments, with state police providing broader services. Federal agencies such as the Federal Bureau of Investigation (FBI) and the U.S. Marshals Service have specialized duties.[229] At the federal level and in almost every state, jurisprudence operates on a common law system.

State courts conduct most criminal trials; federal courts handle certain designated crimes as well as certain appeals from the state criminal courts. Federal law prohibits a variety of drugs, although states sometimes pass laws in conflict with federal regulations. The smoking age is generally 18, and the drinking age is generally 21.

Among developed nations, the United States has above-average levels of violent crime and particularly high levels of gun violence and homicide.[230] There were 4.7 murders per 100,000 persons in 2011, 14.5% fewer than in 2000 (5.5), and 19.0% fewer since a recent peak of 5.8 in 2006.[] [] Gun ownership rights are the subject of contentious political debate.

The United States has the highest documented incarceration rate[] and total prison population[231] in the world. At the start of 2008, more than 2.3 million people were incarcerated, more than one in every 100 adults.[232] The current rate is about seven times the 1980 figure,[233] and over three times the figure in Poland, the Organisation for Economic Co-operation and Development (OECD) country with the next highest rate.[234] African American males are jailed at about six times the rate of white males and three times the rate of Hispanic males.[] The country's high rate of incarceration is largely due to sentencing and drug policies.[] []

Capital punishment is sanctioned in the United States for certain federal and military crimes, and used in 32 states.[235] While there are 32 states which include capital punishment within their sentencing statutes, some states (such as New Hampshire and Kansas) have yet to execute anyone since 1976, as demonstrated by the lack of any executions by these states out of the 1317 total executions which have taken place as of December 5, 2012 [236] No executions took place from 1967 to 1977, owing in part to a U.S. Supreme Court ruling striking down arbitrary imposition of the death penalty. In 1976, that Court ruled that, under appropriate circumstances, capital punishment may constitutionally be imposed; since the decision there have been more than 1,300 executions, a majority of these taking place in three states: Texas, Virginia, and Oklahoma.[237] Five state legislatures in the modern era have abolished the death penalty, though two of those laws (in New Mexico and Connecticut) were not retroactive. Additionally, state courts in Massachusetts and New York struck down death penalty statutes and their legislatures took no action in response. In 2010, the country had the fifth highest number of executions in the world, following China, Iran, North Korea, and Yemen.[238]

Economy

Economic indicators		
Nominal GDP	15.984 trillion (Q1 2013)	[]
Real GDP growth	1.8% (Q1 2013, annualized)	
	2.2% (2012)	[239]
CPI inflation	2.0% (February 2012 – February 2013)	[240]
Employment-to-population ratio	58.5% (March 2013)	[241]
Unemployment	7.4% (July 2013)	[242]
Labor force participation rate	63.3% (March 2013)	[243]
Poverty	15.1% (2010)	[]
Public debt	$16.433 trillion (Q4 2012)	[244]
Household net worth	$58.5 trillion (Q4 2011)	[245]

The United States has a capitalist mixed economy, which is fueled by abundant natural resources, a well-developed infrastructure, and high productivity.[246] According to the International Monetary Fund, the U.S. GDP of $15.1 trillion constitutes 22% of the gross world product at market exchange rates and over 19% of the gross world product at purchasing power parity (PPP).[] Though larger than any other nations, its national GDP was about 5% smaller at PPP in 2011 than the European Union's, whose population is around 62% higher.[247] The country ranks ninth in the world in nominal GDP per capita and sixth in GDP per capita at PPP.[] The U.S. dollar is the world's primary reserve currency.[248]

The United States is the largest importer of goods and second largest exporter, though exports per capita are relatively low. In 2010, the total U.S. trade deficit was $635 billion.[] Canada, China, Mexico, Japan, and Germany are its top trading partners.[249] In 2010, oil was the largest import commodity, while transportation equipment was the country's largest export.[] China is the largest foreign holder of U.S. public debt.[250]

The New York Stock Exchange on Wall Street, the world's largest stock exchange per total market capitalization of its listed companies.[251]

In 2009, the private sector was estimated to constitute 86.4% of the economy, with federal government activity accounting for 4.3% and state and local government activity (including federal transfers) the remaining 9.3%.[252] While its economy has reached a postindustrial level of development and its service sector constitutes 67.8% of GDP, the United States remains an industrial power.[] The leading business field by gross business receipts is wholesale and retail trade; by net income it is manufacturing.[253]

Chemical products are the leading manufacturing field.[254] The United States is the third largest producer of oil in the world, as well as its largest importer.[255] It is the world's number one producer of electrical and nuclear energy, as well as liquid natural gas, sulfur, phosphates, and salt. While agriculture accounts for just under 1% of GDP,[] the United States is the world's top producer of corn[256] and soybeans.[257] Coca-Cola and McDonald's are the two most recognized brands in the world.[258]

Consumer spending comprises 71% of the US economy in 2013.[259] In August 2010, the American labor force consisted of 154.1 million people. With 21.2 million people, government is the leading field of employment. The largest private employment sector is health care and social assistance, with 16.4 million people. About 12% of workers are unionized, compared to 30% in Western Europe.[260] The World Bank ranks the United States first in the ease of hiring and firing workers.[] In 2009, the United States had the third highest labor productivity per person in the world, behind Luxembourg and Norway. It was fourth in productivity per hour, behind those two countries and the Netherlands.[261]

The 2008-2012 global recession had a significant impact on the United States. For example, persistent high unemployment remains, along with low consumer confidence, the continuing decline in home values and increase in foreclosures and personal bankruptcies, an escalating federal debt crisis, inflation, and rising petroleum and food prices. In fact, a 2011 poll found that more than half of all Americans think the U.S. is still in recession or even depression, despite official data that shows a historically modest recovery.[262]

Income, poverty, and wealth

Americans have the highest average household and employee income among OECD nations, and, as of 2007, the second highest median household income, behind only Luxembourg.[] [] According to the Census Bureau real median household income was $50,502 in 2011, down from $51,144 in 2010.[263] The Global Food Security Index ranked the US #1 in food affordability and overall food security in March 2013.[264] On average Americans have over twice as much living space per dwelling and per person as Europeans, and more than every European nation.[]

The US economy is currently embroiled in the economic downturn which followed the Financial crisis of 2007–2008, with output still below potential according to the CBO[265] and unemployment still above historic trends.[266] As of February 2013, the unemployment rate was 7.7% or 12.0 million people, while the government's broader U-6 unemployment rate, which includes the part-time underemployed was 14.3% or 22.2 million. With a record proportion of long term unemployed, continued decreasing household income, tax hikes, and new federal budget cuts, the US economy remained in a jobless recovery.[] [267] Half of the U.S. population lives in poverty or is low-income, according to U.S. census data.[268] According to a survey by the Associated Press, four out of five U.S. adults struggle with joblessness, near-poverty or reliance on welfare for at least parts of their lives.[269]

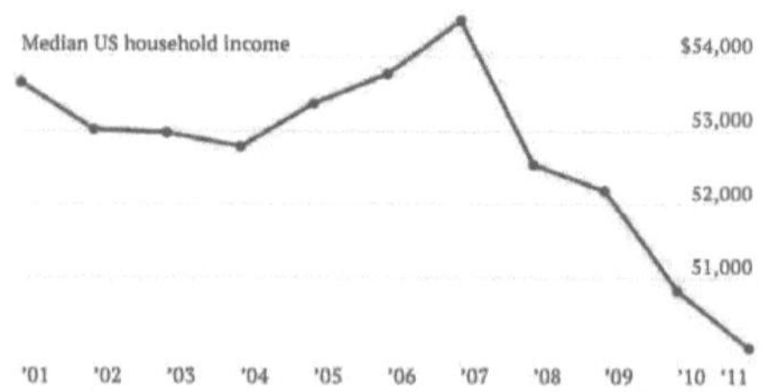

The median US household income, 2001–2011.

While inflation-adjusted ("real") household income had been increasing almost every year from 1947 to 1999, it has since been flat and even decreased recently.[270] Extreme poverty in the United States, meaning households living on less than $2 per day before government benefits, doubled from 1996 levels to 1.5 million households in 2011, including 2.8 million children.[271] In 2011 16.7 million children lived in food insecure households, about 35% more than 2007 levels, though only 1.1% of US children, or 845,000, saw reduced food intake or disrupted eating patterns at some point during the year, and most cases weren't chronic.[272]

There were about 643,000 sheltered and unsheltered homeless persons in the U.S. in January 2009. Almost two-thirds stayed in an emergency shelter or transitional housing program and the other third were living on the street, in an abandoned building, or another place not meant for human habitation. About 1.56 million people, or about 0.5% of the U.S. population, used an emergency shelter or a transitional housing program between October 1, 2008 and September 30, 2009.[] The U.S. welfare state is one of the least extensive in the developed world, reducing both relative poverty and absolute poverty by considerably less than the mean for rich nations,[] [273] [274] though combined private and public social expenditures per capita are relatively high and Americans face much lower consumption taxes than poor Europeans.[275]

While the American welfare state effectively reduces poverty among the elderly,[276] it provides relatively little assistance to the young.[277] A 2007 UNICEF study of children's well-being in 21 industrialized nations, based on factors like income relative to each nation's own median, self-reported risky behavior, and family relationship quality, ranked the United States next to last.[278] The share of income of the top 1%—21.8% of total reported income in 2005—has more than doubled since 1980,[279] leaving the United States with the greatest income inequality among developed nations.[] [280]

Poverty in the US has been increasing as median incomes have declined. Analyses using a common data set for comparisons tend to find that the US has a lower absolute poverty rate by market income than most other wealthy nations.[281] Over 80% of poor American households have air conditioning, three quarters own at least one automobile, about 40% own their homes, and the average poor American has more living space than the general population average in every European nation except Luxembourg and Denmark. Most of them have a refrigerator, stove, microwave, telephone, and television. About half have computers and less than half have internet service.[]

Wealth, like income and taxes, is highly concentrated: The richest 10% of the adult population possesses 69.8% of the country's household wealth, the second-highest share among developed nations.[282] In 2013 the United Nations Development Programme ranked the United States 16th among 132 countries on its inequality-adjusted human development index (IHDI), 13 places lower than in the standard HDI.[283] For the year 2012, the United States ranks 12th on the Legatum Prosperity Index[284] and 19th on the Corruption Perceptions Index.[285]

A middle-class suburban housing development in San Jose, California

Between June 2007 and November 2008 the global recession led to falling asset prices around the world. Assets owned by Americans lost about a quarter of their value.[286] Since peaking in the second quarter of 2007, household wealth is down $14 trillion.[287] At the end of 2008, household debt amounted to $13.8 trillion.[288] By some measures, the US has more millionaires per capita than any other nation, ranks in the top 14 in billionaires per capita,[289] and has more billionaires and millionaires than any other nation and all of Europe; most of them self-made.[290] [291] [292]

Infrastructure

Transportation

Personal transportation is dominated by automobiles, which operate on a network of 13 million roads, including one of the world's longest highway systems.[294] The world's second largest automobile market,[295] the United States has the highest rate of per-capita vehicle ownership in the world, with 765 vehicles per 1,000 Americans.[296] About 40% of personal vehicles are vans, SUVs, or light trucks.[297] The average American adult (accounting for all drivers and non-drivers) spends 55 minutes driving every day, traveling 29 miles (km).[298]

The Interstate Highway System, which extends 46876 miles (km)[293]

Mass transit accounts for 9% of total U.S. work trips.[299] [300] While transport of goods by rail is extensive, relatively few people use rail to travel,[301] though ridership on Amtrak, the national intercity passenger rail system, grew by almost 37% between 2000 and 2010.[302] Also, light rail development has increased in recent years.[303] Bicycle usage for work commutes is minimal.[304]

The civil airline industry is entirely privately owned and has been largely deregulated since 1978, while most major airports are publicly owned. The three largest airlines in the world by passengers carried are U.S.-based; American Airlines is number one after its 2013 acquisition of US Airways.[305] Of the world's 30 busiest passenger airports, 16 are in the United States, including the busiest, Hartsfield-Jackson Atlanta International Airport.[306]

Further information: List of airlines of the United States

Energy

The United States energy market is 29,000 terawatt hours per year. Energy consumption per capita is 7.8 tons of oil equivalent per year, the 10th highest rate in the world. In 2005, 40% of this energy came from petroleum, 23% from coal, and 22% from natural gas. The remainder was supplied by nuclear power and renewable energy sources.[307] The United States is the world's largest consumer of petroleum.[308]

For decades, nuclear power has played a limited role relative to many other developed countries, in part because of public perception in the wake of a 1979 accident. In 2007, several applications for new nuclear plants were filed.[309] The United States has 27% of global coal reserves.[] It is the world's largest producer of natural gas and crude oil.[310]

Science and technology

Neil Armstrong was the first person to walk on the Moon.

The United States has been a leader in scientific research and technological innovation since the late 19th century. In 1876, Alexander Graham Bell was awarded the first U.S. patent for the telephone. Thomas Edison's laboratory developed the phonograph, the first long-lasting light bulb, and the first viable movie camera.[311] Nikola Tesla pioneered alternating current, the AC motor, and radio. In the early 20th century, the automobile companies of Ransom E. Olds and Henry Ford popularized the assembly line. The Wright brothers, in 1903, made the first sustained and controlled heavier-than-air powered flight.[312]

The rise of Nazism in the 1930s led many European scientists, including Albert Einstein, Enrico Fermi, and John von Neumann, to immigrate to the United States. During World War II, the Manhattan Project developed nuclear weapons, ushering in the Atomic Age. The Space Race produced rapid advances in rocketry, materials science, and computers. IBM, Apple Computer, and Microsoft refined and popularized the personal computer.

The United States largely developed the ARPANET and its successor, the Internet. Today, 64% of research and development funding comes from the private sector.[313] The United States leads the world in scientific research papers and impact factor.[314] As of April 2010, 77% of American households owned at least one computer, and 68% had broadband Internet service.[315] 85% of Americans also own a mobile phone as of 2011.[316] The country is the primary developer and grower of genetically modified food, representing half of the world's biotech crops.[317]

Education

Some 80% of U.S. college students attend public universities such as the University of Virginia, founded by Thomas Jefferson[318]

American public education is operated by state and local governments, regulated by the United States Department of Education through restrictions on federal grants. In most states, children are required to attend school from the age of six or seven (generally, kindergarten or first grade) until they turn 18 (generally bringing them through twelfth grade, the end of high school); some states allow students to leave school at 16 or 17.[319] About 12% of children are enrolled in parochial or nonsectarian private schools. Just over 2% of children are homeschooled.[320]

The United States has many competitive private and public institutions of higher education. According to prominent international rankings, 13 or 15 American colleges and universities are ranked among the top 20 in the world.[321] [322] There are also local community colleges with generally more open admission policies, shorter academic programs, and lower tuition. Of Americans 25 and older, 84.6% graduated from high school, 52.6% attended some college, 27.2% earned a bachelor's degree, and 9.6% earned graduate degrees.[323] The basic literacy rate is approximately 99%.[] [324] The United Nations assigns the United States an Education Index of 0.97, tying it for 12th in the world.[325]

Health

The Texas Medical Center in Houston is the world's largest medical center.

The United States has life expectancy of 78.4 years at birth, up from 75.2 years in 1990, ranks it 50th among 221 nations, and 27th out of the 34 industrialized OECD countries, down from 20th in 1990.[326] [] Increasing obesity in the United States and health improvements elsewhere have contributed to lowering the country's rank in life expectancy from 1987, when it was 11th in the world.[327] Obesity rates in the United States are among the highest in the world.[328] Approximately one-third of the adult population is obese and an additional third is overweight;[329] the obesity rate, the highest in the industrialized world, has more than doubled in the last quarter-century.[330] Obesity-related type 2 diabetes is considered epidemic by health care professionals.[331] The infant mortality rate of 6.06 per thousand places the United States 176th highest out of 222 countries.[332]

In 2010, coronary artery disease, lung cancer, stroke, chronic obstructive pulmonary diseases, and traffic accidents caused the most years of life lost in the US. Low back pain, depression, musculoskeletal disorders, neck pain, and anxiety caused the most years lost to disability. The most deleterious risk factors were poor diet, tobacco smoking, obesity, high blood pressure, high blood sugar, physical inactivity, and alcohol use. Alzheimer's disease, drug abuse, kidney disease and cancer, and falls caused the most additional years of life lost over their age-adjusted 1990 per-capita rates.[]

The U.S. is a global leader in medical innovation. America solely developed or contributed significantly to 9 of the top 10 most important medical innovations since 1975 as ranked by a 2001 poll of physicians, while the EU and Switzerland together contributed to five. Since 1966 Americans have received more Nobel Prizes in Medicine than the rest of the world combined. From 1989 to 2002 four times more money was invested in private biotechnology companies in America than in Europe.[333] [334]

The U.S. health-care system far outspends any other nations, measured in both per capita spending and percentage of GDP.[335] In 2008, the US spent more on health care per capita ($7,146), and as percentage of GDP (15.2%), than any other nation. Health-care coverage in the United States is a combination of public and private efforts, and is not universal as in all other developed countries. In 2004, private insurance paid for 36% of personal health expenditures, private out-of-pocket payments covered 15%, and federal, state, and local governments paid for 44%.[]

In 2010, 49.9 million residents or 16.3% of the population did not carry health insurance. The main cause of this rise is the drop in the number of Americans with employer-sponsored health insurance.[] The subject of uninsured and underinsured Americans is a major political issue.[336] [337] In 2006, Massachusetts became the first state to mandate universal health insurance.[338] In 2007, 62.1% of filers for bankruptcy blamed medical expenses. About 25% of all senior citizens declare bankruptcy because of medical expenses, and 43% are forced to mortgage or sell their primary residence.[339] Federal legislation passed in early 2010 would ostensibly create a near-universal health insurance system around the country by 2014, though the bill and its ultimate impact are issues of controversy.[340] [341]

Culture

The Statue of Liberty in New York City is a symbol of both the U.S. and ideals of freedom, democracy, and opportunity.[342]

The United States is a multicultural nation, home to a wide variety of ethnic groups, traditions, and values.[8] [343] Aside from the relatively small Native American and Native Hawaiian populations, nearly all Americans or their ancestors settled or immigrated within the past five centuries.[344] Mainstream American culture is a Western culture largely derived from the traditions of European immigrants with influences from many other sources, such as traditions brought by slaves from Africa.[8] [345] More recent immigration from Asia and especially Latin America has added to a cultural mix that has been described as both a homogenizing melting pot, and a heterogeneous salad bowl in which immigrants and their descendants retain distinctive cultural characteristics.[8]

American culture is considered the most individualistic in the world.[346] The American Dream, or the perception that Americans enjoy high social mobility, plays a key role in attracting immigrants.[347] By some measures, social mobility is actually lower than other high-income countries such as France, Germany, Canada, Australia, and the Nordic countries,[348] though such studies are based on relative comparisons within each nation rather than absolute wealth earned throughout one's life, the US having both a more stretched-out income distribution and a higher median income than those nations.[347] [349] While the mainstream culture holds that the United States is a classless society,[350] scholars identify significant differences between the country's social classes, affecting socialization, language, and values.[351]

The American middle and professional class has initiated many contemporary social trends such as modern feminism, environmentalism, and multiculturalism.[352] Americans' self-images, social viewpoints, and cultural expectations are associated with their occupations to an unusually close degree.[353] While Americans tend greatly to value socioeconomic achievement, being ordinary or average is generally seen as a positive attribute.[354]

Popular media

The world's first commercial motion picture exhibition was given in New York City in 1894, using Thomas Edison's Kinetoscope. The next year saw the first commercial screening of a projected film, also in New York, and the United States was in the forefront of sound film's development in the following decades. Since the early 20th century, the U.S. film industry has largely been based in and around Hollywood, California.

Director D. W. Griffith was central to the development of film grammar and Orson Welles's *Citizen Kane* (1941) is frequently cited as the greatest film of all time.[355] [356] American screen actors like John Wayne and Marilyn Monroe have become iconic figures, while producer/entrepreneur Walt Disney was a leader in both animated film and movie merchandising. Hollywood is also one of the leaders in motion picture production.[357]

Americans are the heaviest television viewers in the world,[358] and the average viewing time continues to rise, reaching five hours a day in 2006.[359] The four major broadcast television networks are all commercial entities. Americans listen to radio programming, also largely commercialized, on average just over two-and-a-half hours a day.[360] Aside from web portals and search engines, the most popular websites are Facebook, YouTube, Wikipedia, Blogger, eBay, and Craigslist.[]

The rhythmic and lyrical styles of African-American music have deeply influenced American music at large, distinguishing it from European traditions. Elements from folk idioms such as the blues and what is now known as old-time music were adopted and transformed into popular genres with global audiences. Jazz was developed by innovators such as Louis Armstrong and Duke Ellington early in the 20th century. Country music developed in the 1920s, and rhythm and blues in the 1940s.[361]

Elvis Presley and Chuck Berry were among the mid-1950s pioneers of rock and roll. In the 1960s, Bob Dylan emerged from the folk revival to become one of America's most celebrated songwriters and James Brown led the development of funk. More recent American creations include hip hop and house music. American pop stars such as Presley, Michael Jackson, and Madonna have become global celebrities.[361]

Literature, philosophy, and the arts

In the 18th and early 19th centuries, American art and literature took most of its cues from Europe. Writers such as Nathaniel Hawthorne, Edgar Allan Poe, and Henry David Thoreau established a distinctive American literary voice by the middle of the 19th century. Mark Twain and poet Walt Whitman were major figures in the century's second half; Emily Dickinson, virtually unknown during her lifetime, is now recognized as an essential American poet.[362] A work seen as capturing fundamental aspects of the national experience and character—such as Herman Melville's *Moby-Dick* (1851), Twain's *The Adventures of Huckleberry Finn* (1885), and F. Scott Fitzgerald's *The Great Gatsby* (1925)—may be dubbed the "Great American Novel".[363]

Mark Twain, famous American author and humorist.

Eleven U.S. citizens have won the Nobel Prize in Literature, most recently Toni Morrison in 1993. William Faulkner and Ernest Hemingway are often named among the most influential writers of the 20th century.[364] Popular literary genres such as the Western and hardboiled crime fiction developed in the United States. The Beat Generation writers opened up new literary approaches, as have postmodernist authors such as John Barth, Thomas Pynchon, and Don DeLillo.

The transcendentalists, led by Thoreau and Ralph Waldo Emerson, established the first major American philosophical movement. After the Civil War, Charles Sanders Peirce and then William James and John Dewey were leaders in the development of pragmatism. In the 20th century, the work of W. V. O. Quine and Richard Rorty, and later Noam Chomsky, brought analytic philosophy to the fore of American philosophical academia. John Rawls and Robert Nozick led a revival of political philosophy. Cornel West and Judith Butler have led a continental tradition in American philosophical academia. Globally influential Chicago school economists like Milton Friedman, James M. Buchanan, and Thomas Sowell have transcended discipline to impact various fields in social and political philosophy.[365] [366]

In the visual arts, the Hudson River School was a mid-19th-century movement in the tradition of European naturalism. The realist paintings of Thomas Eakins are now widely celebrated. The 1913 Armory Show in New York City, an exhibition of European modernist art, shocked the public and transformed the U.S. art scene.[367] Georgia O'Keeffe, Marsden Hartley, and others experimented with new, individualistic styles. Major artistic movements such as the abstract expressionism of Jackson Pollock and Willem de Kooning and the pop art of Andy Warhol and Roy Lichtenstein developed largely in the United States. The tide of modernism and then postmodernism has brought fame to American architects such as Frank Lloyd Wright, Philip Johnson, and Frank Gehry.

One of the first major promoters of American theater was impresario P. T. Barnum, who began operating a lower Manhattan entertainment complex in 1841. The team of Harrigan and Hart produced a series of popular musical comedies in New York starting in the late 1870s. In the 20th century, the modern musical form emerged on Broadway; the songs of musical theater composers such as Irving Berlin, Cole Porter, and Stephen Sondheim have become pop standards. Playwright Eugene O'Neill won the Nobel literature prize in 1936; other acclaimed U.S. dramatists include multiple Pulitzer Prize winners Tennessee Williams, Edward Albee, and August Wilson.

Times Square in New York City, the hub of the Broadway Theater District.

Though little known at the time, Charles Ives's work of the 1910s established him as the first major U.S. composer in the classical tradition, while experimentalists such as Henry Cowell and John Cage created a distinctive American approach to classical composition. Aaron Copland and George Gershwin developed a new synthesis of popular and classical music. Choreographers Isadora Duncan and Martha Graham helped create modern dance, while George Balanchine and Jerome Robbins were leaders in 20th-century ballet. Americans have long been important in the modern artistic medium of photography, with major photographers including Alfred Stieglitz, Edward Steichen, and Ansel Adams. The newspaper comic strip and the comic book are both U.S. innovations. Superman, the quintessential comic book superhero, has become an American icon.[368]

Food

Mainstream American cuisine is similar to that in other Western countries. Wheat is the primary cereal grain. Traditional American cuisine uses indigenous ingredients, such as turkey, venison, potatoes, sweet potatoes, corn, squash, and maple syrup, which were consumed by Native Americans and early European settlers.

An apple pie, surrounded by some other icons of American culture.

Slow-cooked pork and beef barbecue, crab cakes, potato chips, and chocolate chip cookies are distinctively American foods. Soul food, developed by African slaves, is popular around the South and among many African Americans elsewhere. Syncretic cuisines such as Louisiana Creole, Cajun, and Tex-Mex are regionally important.

Characteristic dishes such as apple pie, fried chicken, pizza, hamburgers, and hot dogs derive from the recipes of various immigrants. French fries, Mexican dishes such as burritos and tacos, and pasta dishes freely adapted from Italian sources are widely consumed.[] Americans generally prefer coffee to tea. Marketing by U.S. industries is largely responsible for making orange juice and milk ubiquitous breakfast beverages.[369] [370]

The American fast food industry, the world's largest, pioneered the drive-through format in the 1930s. Fast food consumption has sparked health concerns. During the 1980s and 1990s, Americans' caloric intake rose 24%;[] frequent dining at fast food outlets is associated with what public health officials call the American "obesity epidemic".[371] Highly sweetened soft drinks are widely popular, and sugared beverages account for nine percent of American caloric intake.[372]

Sports

Swimmer Michael Phelps is the most decorated Olympic athlete of all time.

The market for professional sports in the United States is roughly $69 billion, roughly 50% larger than that of all of Europe, the Middle East, and Africa combined.[373] Baseball has been regarded as the national sport since the late 19th century, while American football is now by several measures the most popular spectator sport.[374] Basketball and ice hockey are the country's next two leading professional team sports. These four major sports, when played professionally, each occupy a season at different, but overlapping, times of the year. College football and basketball attract large audiences. Boxing and horse racing were once the most watched individual sports,[375] but they have been eclipsed by golf and auto racing, particularly NASCAR. In the 21st century, televised mixed martial arts has also gained a strong following of regular viewers.[376] While soccer is much less popular in the United States than in many other nations, it is played widely at the youth and amateur levels; even so, the men's team has been to the past six World Cups and the women are #1 in the women's world rankings. Tennis and many outdoor sports are popular as well.

While most major U.S. sports have evolved out of European practices, basketball, volleyball, skateboarding, snowboarding, and cheerleading are American inventions, which have become popular in other countries. Lacrosse and surfing arose from Native American and Native Hawaiian activities that predate Western contact. Eight Olympic Games have taken place in the United States. The United States has won 2,301 medals at the Summer Olympic Games, more than any other country,[377] and 253 in the Winter Olympic Games, the second most by 2006.[378]

See also

- Index of United States-related articles
- Outline of the United States
- List of United States cities by population
- List of metropolitan areas of the United States
- List of official United States national symbols
- United States at Wikipedia books

References

[1] 36 U.S.C. § 302 (http://www.law.cornell.edu/uscode/36/302.html) *National motto*

[2] Simonson, 2010

[3] Dept. of Treasury, 2011

[4] (figure updated automatically).

[8] Adams, J.Q.; Strother-Adams, Pearlie (2001). *Dealing with Diversity*. Chicago: Kendall/Hunt. ISBN 0-7872-8145-X.

[9] Greene, Jack P.; Pole, J.R., eds. (2008). *A Companion to the American Revolution*. pp. 352–361.

[11]

[13] The European Union has a larger collective economy, but is not a single nation.

[15] (http://www.infrastructurereportcard.org/,2009)

[19] Cohen, 2004:History and the Hyperpower

[20] BBC, April 2008:Country Profile: United States of America

[22] DeLear, Byron (July 4, 2013) Who coined 'United States of America'? Mystery might have intriguing answer. (http://www.csmonitor.com/USA/Politics/2013/0704/Who-coined-United-States-of-America-Mystery-might-have-intriguing-answer) "Historians have long tried to pinpoint exactly when the name 'United States of America' was first used and by whom. A new find suggests the man might have been George Washington himself." *Christian Science Monitor* (Boston, MA).

[23] *"To the inhabitants of Virginia," by A PLANTER.* Dixon and Hunter's Virginia Gazette #1287 (http://research.history.org/DigitalLibrary/VirginiaGazette/VGIssueThumbs.cfm?IssueIDNo=76.DH.16) – April 6, 1776, Williamsburg, Virginia. Letter is also included in Peter Force's *American Archives* Vol. 5

[24] Carter, Rusty (August 18, 2012). "You read it here first" (http://www.vagazette.com/news/va-vg-usa-first-0818-20120818,0,4983868.story). *Virginia Gazette*. "He did a search of the archives and found the letter on the front page of the April 6, 1776, edition, published by Hunter & Dixon."

[25] DeLear, Byron (August 16, 2012). "Who coined the name 'United States of America'? Mystery gets new twist." (http://www.csmonitor.com/USA/Politics/The-Vote/2012/0816/Who-coined-the-name-United-States-of-America-Mystery-gets-new-twist#disqus_thread) *Christian Science Monitor* (Boston, MA).

[26] Jefferson's "original Rough draught" of the Declaration of Independence (http://www.princeton.edu/~tjpapers/declaration/declaration.html)

[30] Wilson, Kenneth G. (1993). *The Columbia Guide to Standard American English*. New York: Columbia University Press, pp. 27–28. ISBN 0-231-06989-8.

[31] G. H. Emerson, The Universalist Quarterly and General Review, Vol. 28 (Jan. 1891), p. 49, quoted in Zimmer paper above.

[32] For example, the U.S. embassy in Spain calls itself the embassy of the "Estados Unidos", literally the words "states" and "united", and also uses the initials "EE.UU.", the doubled letters implying plural use in Spanish (http://spanish.madrid.usembassy.gov/) Elsewhere on the site "Estados Unidos de América" is used (http://spanish.madrid.usembassy.gov/es/educacion/benjamin_2013.html)

[33] Smithsonian, 2004, Human Origins

[34] " *The Cambridge encyclopedia of human paleopathology* (http://books.google.com/books?id=qubTdDk1H3IC&pg=PA205)". Arthur C. Aufderheide, Conrado Rodríguez-Martín, Odin Langsjoen (1998). Cambridge University Press. p.205. ISBN 0-521-55203-6

[35] Bianchine, Russo, 1992 pp.225-232

[36] Mann, 2005 p.44

[37] Thornton, 1987 p.49

[38] Kessel, 2005 pp.142-143

[39] Mercer Country Historical Society, 2005

[40] Juergens, 2011, p.69

[41] Ripper, 2008 p.6

[42] Ripper, 2008 p.5

[43] Calloway, 1998, p.55

[44] Vaughan, 1999, p.12

[45] Ranlet, 1999, pp.140-141

[46] Ranlet, 1999, p.137

[47] Rausch, 1994, p.59

[48] Sturm, Circe. "Blood Politics, Racial Classification, and Cherokee National Identity: The Trials and Tribulations of the Cherokee Freedmen", *American Indian Quarterly*, Vol. 22, No. 1/2. (Winter – Spring, 1998), p.231.

[49] Taylor, pp. 33-34

[50] Taylor, pp. 72, 74

[51] Walton, 2009, pp. 29-31

[52] Remini, 2007, pp. 2,3,

[53] Johnson, 1997, pp. 26-30

[54] Berkin, 2007, p. 75

[55] Price, 2003

[56] Vaughan, 1999

[57] Walton, 2009, chapter 3

[58] Lemon, 1987

[59] Gordon, 2004, intro, chapters 1 and 2

[60] Russell, David Lee (2005). *The American Revolution in the Southern Colonies*. Jefferson, N.C., and London: McFarland, p. 12. ISBN 0-7864-0783-2.

[61] Clingan, 2000, p.13

[62] Tadman, 2000, p.1534

[63] Schneider, 2007, p.484

[64] Lien, 1913, p.522

[65] Davis, 1996, p.7

[66] Quirk, 2011, p.195

[67] Gold, 2006, pp.32-35

[68] Walton, 2009, pp. 38,39

[69] Walton, 2009, p. 35

[70] Greene and Pole, *A Companion to the American Revolution* p 357
Jonathan R. Dull, *A Diplomatic History of the American Revolution* (1987) p. 161

Lawrence S. Kaplan, "The Treaty of Paris, 1783: A Historiographical Challenge," *International History Review,* Sept 1983, Vol. 5 Issue 3, pp 431–442

[71] Boyer, 2007, pp.192-193

[72] Walton, 2009, p. 43

[73] Gordon, 2004, pp. 27,29

[75] Smith (2001), *Grant*, pp. 525–526

[76] Carlisle (2008), *The Civil War and Reconstruction*, p. 1

[77] Page 7 lists a total slave population of 3,953,760.

[78] De Rosa, Marshall L. (1997). *The Politics of Dissolution: The Quest for a National Identity and the American Civil War*. Edison, NJ: Transaction. p. 266. ISBN 1-56000-349-9.

[79] Vinovskis, Maris (1990). " *Toward a social history of the American Civil War: exploratory essays* (http://books.google.com/books?id=gySktxKYPGoC&pg=PA6)". Cambridge University Press. p. 6. ISBN 0-521-39559-3.

[80] Brands (2012), *Grant Takes On The Klan*, **American History**, p. 46

[83] Zinn, Howard. *A People's History of the United States*. New York: Harper Perennial Modern Classics, 2005. Chapter 13, *"The Socialist Challange"* pp. 321-357 ISBN 0060838655

[84] Zinn, Howard. *A People's History of the United States*. New York: Harper Perennial Modern Classics, 2005. p. 327 ISBN 0060838655

[85] Foner, Eric; Garraty, John A. (1991). *The Reader's Companion to American History*. New York: Houghton Mifflin. p. 576. ISBN 0-395-51372-3.

[86] McDuffie, Jerome; Piggrem, Gary Wayne; Woodworth, Steven E. (2005). *U.S. History Super Review*. Piscataway, NJ: Research & Education Association. p. 418. ISBN 0-7386-0070-9.

[89] See, for example, the aircraft production figures cited in : in 1940, the Americans produced 6,019 aircraft; in 1941, 19,433; in 1942, almost 48,000; in 1943, an astonishing 85,898. Even more were to come in 1944. For comparison, in 1941, the German war economy could only manage about 12,000 aircraft.

[90] p. 2.

[91] Kennedy, Paul (1989). *The Rise and Fall of the Great Powers*. New York: Vintage. p. 358. ISBN 0-679-72019-7.
Indeed, World War II ushered in the zenith of US power in what came to be called the American Century, as , indicates: "Truman presided over the greatest military and economic power the world had ever known. War production had lifted the United States out of the Great Depression and had inaugurated an era of unimagined prosperity. Gross national product increased by 60 percent during the war, total earnings by 50 percent. Despite social unrest, labor agitation, racial conflict, and teenage vandalism, Americans had more discretionary income than ever before. Simultaneously, the US government had built up the greatest war machine in human history. By the end of 1942, the United States was producing more arms than all the Axis states combined, and, in 1943, it made almost three times more armaments than did the Soviet Union. In 1945, the United States had two-thirds of the world's gold reserves, three-fourths of its invested capital, half of its shipping vessels, and half of its manufacturing capacity. Its GNP was three times that of the Soviet Union and more than five times that of Britain. It was also nearing completion of the atomic bomb, a technological and production feat of huge costs and proportions."

[93] Pacific War Research Society (2006). *Japan's Longest Day*. New York: Oxford University Press. ISBN 4-7700-2887-3.

[97]

[98] Social Security History (http://www.ssa.gov/history/lbjsm.html), the United States Social Security Administration

[99] Soss, 2010, p.277

[100] Fraser, 1989

[101] Ferguson, 1986, pp.43-53

[102] Williams, pp.325–331

[103]

[104]

[105] Hayes, 2009

[106] US History.org, 2013

[107] Burns, 2008

[108] Levy, 1996, p.442

[109] Wallander, 2003

[110]

[112]

[113]

[114]

[115]

[118]

[132] Daynes & Sussman, 2010, pp.3, 72, 74-76, 78

[133] Hays, Samuel P. (2000). *A History of Environmental Politics since 1945*.

[134] Rothman, Hal K. (1998).*The Greening of a Nation? Environmentalism in the United States since 1945*

[135] Turner, James Morton (2012). *The Promise of Wilderness*

[136] http://www.census.gov/prod/cen2010/briefs/c2010br-02.pdf

[140] "U.S. Legal Permanent Residents: 2012" (http://www.dhs.gov/sites/default/files/publications/ois_lpr_fr_2012_2.pdf). Office of Immigration Statistics *Annual Flow Report.*

[144] National Survey of Sexual Health and Behavior (http://www.nationalsexstudy.indiana.edu/). Retrieved January 6, 2013.

[146] U.S. Census Bureau: "U.S. Census Bureau Delivers Final State 2010 Census Population Totals for Legislative Redistricting" (http://www.census.gov/2010census/news/releases/operations/cb11-cn123.html) see custom table, 2nd worksheet

[153] "Language Spoken at Home by the U.S. Population, 2010", American Community Survey, U.S. Census Bureau, in *World Almanac and Book of Facts 2012*, p. 615.

[159] Media, Minorities, and Meaning: A Critical Introduction — Page 88, Debra L. Merskin – 2010

[168] Scheb, John M., and John M. Scheb II (2002). *An Introduction to the American Legal System.* Florence, KY: Delmar, p. 6. ISBN 0-7668-2759-3.

[177] Feldstein, Fabozzi, 2011, p.9

[178] Schultz, 2009, pp.164, 453, 503

[179] Schultz, 2009, p.38

[181] See 8 U.S.C. § 1101 (http://www.law.cornell.edu/uscode/8/1101.html)(a)(36) and 8 U.S.C. (a)(38) U.S. Federal Code, Immigration and Nationality Act. 8 U.S.C. § 1101a (http://www.law.cornell.edu/uscode/8/1101a.html)

[183] US General Accounting Office, U.S. Insular Areas. Application of the U.S. Constitution (http://www.gao.gov/assets/230/224900.pdf). November 1997. p. 9. Appendix I, p. 23-38. Retrieved April 29, 2013.

[231] For the latest data, see For other estimates of the incarceration rate in China and North Korea see

[239] Change is based on chained 2005 dollars. Quarterly growth is expressed as an annualized rate.

[246] Wright, Gavin, and Jesse Czelusta, "Resource-Based Growth Past and Present", in *Natural Resources: Neither Curse Nor Destiny*, ed. Daniel Lederman and William Maloney (World Bank, 2007), p. 185. ISBN 0-8213-6545-2.

[259] "Personal Consumption Expenditures (PCE)/Gross Domestic Product (GDP)" (http://research.stlouisfed.org/fred2/graph/?g=hh3) *FRED Graph*, Federal Reserve Bank of St. Louis

[263] "Household Income for States: 2010 and 2011" (http://www.census.gov/prod/2012pubs/acsbr11-02.pdf) *United States Census, American Community Survey Briefs*, September 2012, Appendix Table 1, page 5

[269] Yen, Hope (28 July 2013). 80 Percent Of U.S. Adults Face Near-Poverty, Unemployment: Survey (http://www.huffingtonpost.com/2013/07/28/poverty-unemployment-rates_n_3666594.html). *The Huffington Post.* Retrieved 28 July 2013.

[271] "Extreme Poverty in the United States, 1996 to 2011" (http://npc.umich.edu/publications/policy_briefs/brief28/policybrief28.pdf) *National Poverty Center*, February 2012

[274] Gould, Elise and Wething, Hilary (24 July 2012). "U.S. poverty rates higher, safety net weaker than in peer countries." (http://www.epi.org/publication/ib339-us-poverty-higher-safety-net-weaker/) *Economic Policy Institute.* Retrieved 15 July 2013.

[276] Orr, D. (November–December 2004). "Social Security Isn't Broken: So Why the Rush to 'Fix' It?" In C. Sturr and R. Vasudevan, eds. (2007). *Current Economic Issues*. Boston: Economic Affairs Bureau.

[284] THE 2012 LEGATUM PROSPERITY INDEX TABLE RANKINGS (http://www.prosperity.com/Ranking.aspx). Legatum Institute, 2012.

[287] " Americans' wealth drops $1.3 trillion (http://money.cnn.com/2009/06/11/news/economy/Americans_wealth_drops/?postversion=2009061113)". CNN. June 11, 2009.

[288] " U.S. household wealth falls $11.2 trillion in 2008 (http://www.reuters.com/article/idUSTRE52B58720090312)". Reuters. March 12, 2009.

[324] For more detail on U.S. literacy, see A First Look at the Literacy of America's Adults in the 21st century (http://nces.ed.gov/NAAL/PDF/2006470.PDF), U.S. Department of Education (2003).

[328] "Slideshow: Most obese countries". (http://www.reuters.com/news/pictures/slideshow?articleId=USRTXT3DK#a=1) *Reuters.* Retrieved November 22, 2012.

[335] *OECD Health Data 2000: A Comparative Analysis of 29 Countries* [CD-ROM] (OECD: Paris, 2000). See also

[339] "Out-of-Pocket Spending in the Last Five Years of Life" (http://link.springer.com/article/10.1007/s11606-012-2199-x) *Journal of General Internal Medicine*, February 2013, Volume 28, Issue 2, pp 304–309

[343] Thompson, William; Hickey, Joseph (2005). *Society in Focus.* Boston: Pearson. ISBN 0-205-41365-X.

[344] Fiorina, Morris P.; Peterson, Paul E. (2000). *The New American Democracy.* London: Longman, p. 97. ISBN 0-321-07058-5.

[345] Holloway, Joseph E. (2005). *Africanisms in American Culture*, 2d ed. Bloomington: Indiana University Press, pp. 18–38. ISBN 0-253-34479-4. Johnson, Fern L. (1999). *Speaking Culturally: Language Diversity in the United States*. Thousand Oaks, Calif., London, and New Delhi: Sage, p. 116. ISBN 0-8039-5912-5.

[348] Gould, Elise (10 October 2012). "U.S. lags behind peer countries in mobility." (http://www.epi.org/publication/usa-lags-peer-countries-mobility/) *Economic Policy Institute.* Retrieved 15 July 2013.

[349] CAP: *Understanding Mobility in America* (http://www.americanprogress.org/issues/2006/04/b1579981.html). April 26, 2006

[355] *Village Voice*: 100 Best Films of the 20th century (2001) (http://www.filmsite.org/villvoice.html). Filmsite.

[356] *Sight and Sound* Top Ten Poll 2002 (http://www.bfi.org.uk/sightandsound/topten/poll/critics-long.html).. British Film Institute. Retrieved June 19, 2007.

[361] Biddle, Julian (2001). *What Was Hot!: Five Decades of Pop Culture in America.* New York: Citadel, p. ix. ISBN 0-8065-2311-5.

[362] Bloom, Harold. 1999. *Emily Dickinson.* Broomall, PA: Chelsea House. p. 9. ISBN 0-7910-5106-4.

[364] Quinn, Edward (2006). *A Dictionary of Literary and Thematic Terms*. Infobase, p. 361. ISBN 0-8160-6243-9. Seed, David (2009). *A Companion to Twentieth-Century United States Fiction*. Chichester, West Sussex: John Wiley and Sons, p. 76. ISBN 1-4051-4691-5. Meyers, Jeffrey (1999). *Hemingway: A Biography*. New York: Da Capo, p. 139. ISBN 0-306-80890-0.
[367] Brown, Milton W. (1988 1963). *The Story of the Armory Show*. New York: Abbeville. ISBN 0-89659-795-4.
[369] Smith, 2004, pp.131-132
[370] Levenstein, 2003, pp.154–55
[371] Boslaugh, Sarah (2010). "Obesity Epidemic", in *Culture Wars: An Encyclopedia of Issues, Viewpoints, and Voices*, ed. Roger Chapman. Armonk, N.Y.: M. E. Sharpe, pp. 413–14. ISBN 978-0-7656-1761-3.
[373] Global sports market to hit $141 billion in 2012 (http://www.reuters.com/article/2008/06/18/us-pwcstudy-idUSN1738075220080618). Reuters. Retrieved on 2013-07-24.
[374] Maccambridge, Michael (2004). *America's Game: The Epic Story of How Pro Football Captured a Nation*. New York: Random House. ISBN 0-375-50454-0.
[378] Norway is first.

Bibliography

- , Book (http://books.google.com/books/about/Making_America_Volume_1_To_1877_A_Histor.html?id=cyEI21RClZkC)
- a
- , Book (http://books.google.com/books?id=9KT3lI76-0cC&source=gbs_navlinks_s)
- , Book (http://books.google.com/books?id=FS3TnrLu7y8C&source=gbs_navlinks_s)
- , Book (http://books.google.com/books?id=edYbAZ7ECEoC&source=gbs_navlinks_s)
- , Book (http://books.google.com/books?id=Dphzw2cbaoQC&vq=1670&source=gbs_navlinks_s)
- , Book (http://books.google.com/books?id=HIOQ1FYHtcYC&source=gbs_navlinks_s)
- , Book (http://books.google.com/books?id=Juc4fb1Fx1cC&source=gbs_navlinks_s)
- , Book (http://books.google.com/books?id=mbV6kPGO4OAC&source=gbs_navlinks_s)
- , Book (http://books.google.com/books/about/An_Empire_of_Wealth.html?id=rmsUs_KDgHAC)
- , eBook version (http://books.google.com/books/about/A_History_of_the_American_People.html?id=RXSVQjz1_tMC)
- , Book (http://books.google.com/books?id=0q2GwESzhUwC&vq=slave&source=gbs_navlinks_s)
- , Book (http://books.google.com/books?id=laxSyAp89G4C&source=gbs_navlinks_s)
- , Book (http://books.google.com/books?id=gvIeAAAAMAAJ&source=gbs_navlinks_s)
- , PDF (http://cascourses.uoregon.edu/geog471/wp-content/uploads/2010/01/Lemon.pdf)
- , E'Book (http://books.google.com/books?id=UYpVAAAAYAAJ&source=gbs_navlinks_s)
- , Book (http://books.google.com/books?id=24mIQuLBuN8C&pg=PA44)
- , Book (http://www.umich.edu/~historyj/docs/2007-fall/Interracial_Marriage_in_Early_America_Mann.pdf)
- eBook version (http://books.google.com/books?id=_EFbS_7fFcYC&printsec=frontcover#v=onepage&q&f=false)
- , Book (http://books.google.com/books?id=qqxK4KlqKYMC&source=gbs_navlinks_s)
- , Book (http://books.google.com/books?id=iyZMeGMgIgEC&vq=triangle&source=gbs_navlinks_s)
- , Book (http://books.google.com/books/about/The_House.html?id=CAM6J6IoQFQC)
- , Book (http://books.google.com/books?id=vX-fYvoAeHwC&source=gbs_navlinks_s)
- , E'Book (http://books.google.com/books?id=G7AJAAAAIAAJ&source=gbs_navlinks_s)
- , Book (http://books.google.com/books?id=QlemwRTsY20C&vq=census&source=gbs_navlinks_s)
- , Book (http://books.google.com/books?id=f7m713xwK58C&vq=twenty+seven+times&source=gbs_navlinks_s)
- , Book (http://books.google.com/books?id=N6lrAmPlbvIC&pg=PA79&dq=#v=onepage&q&f=false)
- , Book (http://books.google.com/?id=JttyjBoyb3AC&lpg=PA12)
- , Article (http://www.jstor.org/discover/10.2307/2652029?uid=3739560&uid=2129&uid=2&uid=70&uid=4&uid=3739256&purchase-type=article&accessType=RR&sid=21102125115943&

showMyJstorPss=false&seq=1&showAccess=true)
- , Book (http://books.google.com/books?id=NPoAQRgkrOcC&vq=1670&source=gbs_navlinks_s)
- , Book (http://books.google.com/?id=9iQYSQ9y60MC&lpg=PA49)
- , Book (http://books.google.com/books/about/History_of_the_American_Economy_With_Acc.html?id=lyhI1q_E4G0C)

Website sources

- , Book (http://books.google.com/books?id=2MjxPJ9W4gwC&source=gbs_navlinks_s)

External links

- United States (https://www.cia.gov/library/publications/the-world-factbook/geos/us.html) entry at *The World Factbook*
- United States (http://www.bbc.co.uk/news/world-us-canada-16761057), from the BBC News
- Key Development Forecasts for the United States (http://www.ifs.du.edu/ifs/frm_CountryProfile.aspx?Country=US) from International Futures

Government

- Official U.S. Government Web Portal (http://www.usa.gov/) Gateway to government sites
- House (http://www.house.gov/) Official site of the United States House of Representatives
- Senate (http://www.senate.gov/) Official site of the United States Senate
- White House (http://www.whitehouse.gov/) Official site of the President of the United States
- Supreme Court (http://www.supremecourt.gov/) Official site of the Supreme Court of the United States

History

- Historical Documents (http://www.nationalcenter.org/HistoricalDocuments.html) Collected by the National Center for Public Policy Research
- U.S. National Mottos: History and Constitutionality (http://www.religioustolerance.org/nat_mott.htm) Analysis by the Ontario Consultants on Religious Tolerance
- USA (http://www.historicalstatistics.org/index2.html) Collected links to historical data

Maps

- National Atlas of the United States (http://nationalatlas.gov/) Official maps from the U.S. Department of the Interior
- Wikimedia Atlas of the United States
- Measure of America (http://www.measureofamerica.org/maps/) A variety of mapped information relating to health, education, income, and demographics for the U.S.

Kentucky

Kentucky /kɪnˈtʌki/, officially the **Commonwealth of Kentucky**, is a state located in the east south-central region of the United States. Kentucky is one of four U.S. states constituted as a commonwealth (the others being Virginia, Pennsylvania, and Massachusetts). Originally a part of Virginia, in 1792 Kentucky became the 15th state to join the Union. Kentucky is the 37th most extensive and the 26th most populous of the 50 United States.

Kentucky is known as the "Bluegrass State", a nickname based on the bluegrass found in many of its pastures because of the fertile soil. One of the major regions in Kentucky is the Bluegrass Region in central Kentucky which houses two of its major cities, Lexington and Louisville. It is a land with diverse environments and abundant resources, including the world's longest cave system, Mammoth Cave National Park, the greatest length of navigable waterways and streams in the contiguous United States, and the two largest man-made lakes east of the Mississippi River.

Kentucky is also home to the highest per capita number of deer and turkey in the United States, the largest free-ranging elk herd east of the Mississippi River, and the nation's most productive coalfield. Kentucky is also known for horse racing, bourbon distilleries, automobile manufacturing, tobacco, bluegrass music, and college basketball.

Etymology

Narrow country roads bounded by stone and wood plank fences are a fixture in the Kentucky Bluegrass region.

It is generally accepted that the historic Native American tribes who hunted in what is now Kentucky referred to the region as *Catawba,* or some similar variant. According to *The Kentucky Blue Book*, Dragging Canoe, a young Cherokee chief opposed to selling ancestral hunting grounds, warned the whites that they were purchasing a "dark and bloody ground."

The origin of Kentucky's modern name (variously spelled *Cane-tuck-ee*, *Cantucky*, *Kain-tuck-ee*, and *Kentuckee* before its modern spelling was accepted)[] comes from an Iroquois word meaning "meadow lands", referring to the buffalo hunting grounds in Central Kentucky's savanna. Members of the *Haudenosaunee*, the Iroquois Confederacy, were historically based in New York and Pennsylvania. They penetrated to this area of the Ohio River Valley and drove other tribes out in order to control more hunting land. In addition to buffalo, they trapped beaver for the lucrative fur trade with the French and English, long before European-American settlement in this area.[1]

Geography

Kentucky is situated in the Upland South.[2] [3] A significant portion of eastern Kentucky is part of Appalachia.

Kentucky borders seven states, from the Midwest and the Southeast. West Virginia lies to the east, Virginia to the southeast, Tennessee to the south, Missouri to the west, Illinois and Indiana to the northwest, and Ohio to the north and northeast. Only Missouri and Tennessee, both of which border eight states, touch more states.

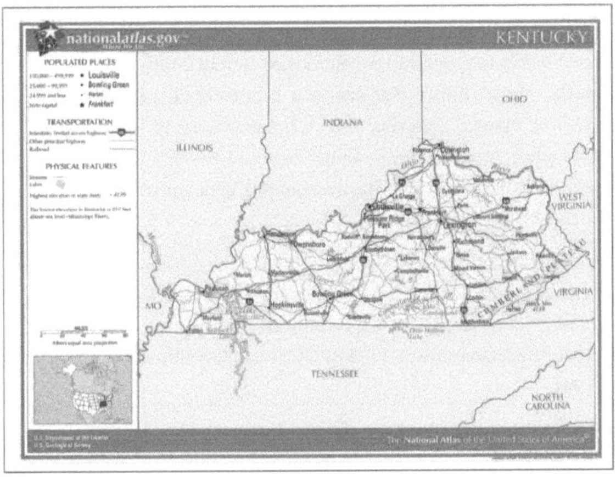

A map of Kentucky

Kentucky's northern border is formed by the Ohio River and its western border by the Mississippi River. The official state borders are based on the courses of the rivers as they existed when Kentucky became a state in 1792. In several places, the rivers have changed courses away from the original borders. For instance, northbound travelers on US 41 from Henderson, after crossing the Ohio River, will be in Kentucky for about two miles. Ellis Park, a thoroughbred racetrack, is located in this small piece of Kentucky. Waterworks Road is part of the only land border between Indiana and Kentucky.[4]

Kentucky is the only U.S. state to have a non-contiguous part existing as an exclave surrounded by other states. Fulton County, in the far west corner of the state, includes Kentucky Bend. This small part of Kentucky on the Mississippi River, bordered by Missouri and accessible via Tennessee, was created by the 1812 New Madrid Earthquake changing the course of the river.[5]

Regions

Kentucky can be divided into five primary regions: the Cumberland Plateau in the east, the north-central Bluegrass region, the south-central and western Pennyroyal Plateau, the Western Coal Fields and the far-west Jackson Purchase.

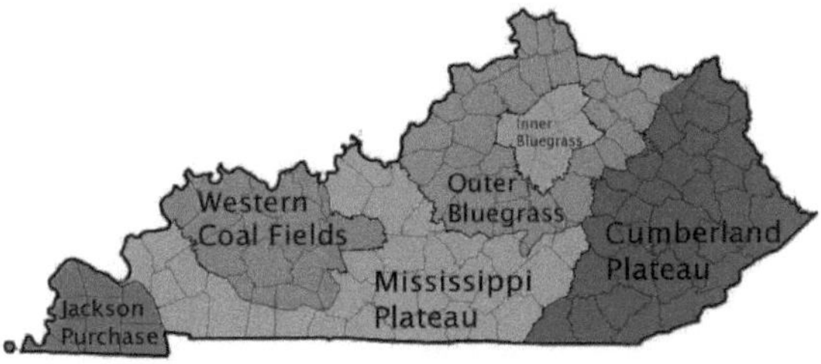

Kentucky's regions (click on image for color-coding information.)

The Bluegrass region is commonly divided into two regions, the Inner Bluegrass—the encircling 90 miles (145 km) around Lexington—and the Outer Bluegrass—the region that contains most of the northern portion of the state, above the Knobs. Much of the outer Bluegrass is in the Eden Shale Hills area, made up of short, steep, and very narrow hills.

This map is a rough depiction of the regions because it relies largely on county lines; as a result, the Inner Bluegrass appears larger than it is, and the Cumberland Plateau appears slightly smaller. The latter region is more commonly known in Kentucky as the East Kentucky Coal Field. Note the singular; these regions are not the sites of coal "fields" but one continuous field with many overlapping seams; the West Kentucky Coal Field is part of the Illinois Basin.

Climate

The East Kentucky Coal Field is known for its rugged terrain.

Located within the southeastern interior portion of North America, Kentucky has a climate that can best be described as a humid subtropical climate (Koppen *Cfa*). Temperatures in Kentucky usually range from daytime summer highs of 87 °F (°C) to the winter low of 23 °F (– °C). The average precipitation is 46 inches (mm) a year.[6]

Kentucky experiences four seasons, with greater variations in the severity of summer and winter.[7] the highest recorded temperature was 114 °F (°C) at Greensburg on July 28, 1930 while the lowest recorded temperature was –37 °F (– °C) at Shelbyville on January 19, 1994.

Major weather events that have affected Kentucky include:

Kentucky's Inner Bluegrass region features hundreds of horse farms.

The Jackson Purchase and western Pennyrile are home to several bald cypress/tupelo swamps.

Event	Death Toll
Louisville Tornado of 1890	est. 76–120+
Ohio River flood of 1937	?
April 3, 1974 Tornado Outbreak	72
April 7, 1977 Flooding (Cumberland River toppled Pineville floodwall)	?
March 1, 1997 Flooding [8]	18
North American blizzard of 2003	?
2008 Super Tuesday tornado outbreak	Weather.com reported 17 deaths
September 2008 Windstorm	1
January 2009 ice storm	24+
March 2012 Tornado Outbreak	22

Monthly Average High and Low Temperatures For Various Kentucky Cities												
City	Jan	Feb	Mar	Apr	May	Jun	Jul	Aug	Sep	Oct	Nov	Dec
Lexington	40/24	45/28	55/36	65/44	74/54	82/62	86/66	85/65	78/58	67/46	54/37	44/28
Louisville	41/25	47/28	57/37	67/46	75/56	83/65	87/70	86/68	79/61	68/48	56/39	45/30
Owensboro	42/25	47/28	57/36	69/45	78/55	86/64	88/68	88/65	81/58	70/46	57/37	45/28
Paducah	42/24	48/28	58/37	68/46	77/55	85/64	89/68	87/65	81/57	71/45	57/36	46/28
Pikeville	46/23	50/25	60/32	69/39	77/49	84/58	87/63	86/62	80/56	71/42	60/33	49/26
Ashland	42/19	47/21	57/29	68/37	77/47	84/56	88/61	87/59	80/52	69/40	57/31	46/23

Lakes and rivers

Kentucky's 90000 miles (km) of streams provides one of the most expansive and complex stream systems in the nation. Kentucky has both the largest artificial lake east of the Mississippi in water volume (Lake Cumberland) and surface area (Kentucky Lake). It is the only U.S. state to be bordered on three sides by rivers—the Mississippi River to the west, the Ohio River to the north, and the Big Sandy River and Tug Fork to the east.[] Its major internal rivers include the Kentucky River, Tennessee River, Cumberland River, Green River and Licking River.

Lake Cumberland is the largest artificial American lake east of the Mississippi River by volume.

Though it has only three major natural lakes,[] the state is home to many artificial lakes. Kentucky also has more navigable miles of water than any other state in the union, other than Alaska.[9]

Natural environment and conservation

Kentucky has an expansive park system which includes one national park, two National Recreation areas, two National Historic Parks, two national forests, two National Wildlife Refuges, 45 state parks, 37696 acres (km^2) of state forest, and 82 Wildlife Management Areas.

Once an industrial wasteland, Louisville's reclaimed waterfront now features thousands of trees and miles of walking trails.

Kentucky has been part of two of the most successful wildlife reintroduction projects in United States history. In the winter of 1997, the Kentucky Department of Fish and Wildlife Resources began to re-stock elk in the state's eastern counties, which had been extinct from the area for over 150 years. As of 2009, the herd had reached the project goal of 10,000 animals, making it the largest herd east of the Mississippi River.[10]

The state also stocked wild turkeys in the 1950s. Once extinct here, more wild turkeys thrive in Kentucky today than in any other eastern state. Hunters telechecked a record 29,006 birds taken during the 23-day season in Spring 2009.[11]

Surrounding states

- Tennessee
- Indiana
- Illinois
- Missouri
- Virginia
- West Virginia
- Ohio

Significant natural attractions

Red River Gorge is one of Kentucky's most visited places

- Cumberland Gap, chief passageway through the Appalachian Mountains in early American history.
- Cumberland Falls, the only place in the Western Hemisphere where a "moon-bow" may be regularly seen, due to the spray of the falls.[12]
- Mammoth Cave National Park, featuring the world's longest known cave system.[13]
- Red River Gorge Geological Area, part of the Daniel Boone National Forest.
- Land Between the Lakes, a National Recreation Area managed by the United States Forest Service.
- Big South Fork National River and Recreation Area near Whitley City.
- Black Mountain, state's highest point.[] Runs along the south ridge of Pine Mountain in Letcher County, Kentucky.The highest point located in Harlan County.
- Bad Branch Falls State Nature Preserve, 2639-acre (km^2) state nature preserve on southern slope of Pine Mountain in Letcher County. Includes one of the largest concentrations of rare and endangered species in the state,[14] as well as a 60-foot (m) waterfall and a Kentucky Wild River.
- Jefferson Memorial Forest, located in the southern fringes of Louisville in the Knobs region, the largest municipally run forest in the United States.[15]
- Lake Cumberland, 1255 miles (km) of shoreline located in South Central Kentucky.
- Natural Bridge, located in Slade, Kentucky Powell County
- Breaks Interstate Park, located in southeastern Pike County, Kentucky and Southwestern Virginia. The Breaks is commonly known as the "Grand Canyon of the South."

History

Abraham Lincoln Birthplace near Hodgenville

What is now Kentucky was inhabited by varying cultures of Native Americans from at least 1000 BC to about AD 1650, particularly along the waterways and in areas of game. Bison roamed in the region. By the time that European and colonial explorers and settlers began entering Kentucky in greater number in the mid-18th century, there were no major Native American settlements in the region. The Iroquois had controlled much of the Ohio River valley for hunting from their bases in what is now New York.

The Shawnee from the northwest and Cherokee from the south also sent parties into the area regularly for hunting. As more settlers entered the area, warfare broke out because the American Indians considered the settlers to be encroaching on their traditional hunting grounds.[16] Today the Southern Cherokee Nation of Kentucky is a state-recognized tribe, although the three federally recognized Cherokee tribes have all vociferously disparaged the authenticity of state-recognized entities claiming to be "Cherokee".[17]

According to a 1790 U.S. government report, 1,500 Kentucky settlers had been killed in Indian raids since the end of the Revolutionary War.[18] In an attempt to end such raids into the state, George Rogers Clark led an expedition of 1,200 drafted men against Shawnee towns on the Wabash River in 1786, one of the first actions of the Northwest Indian War.[19]

After the American Revolution, the counties of Virginia beyond the Appalachian Mountains became known as Kentucky County.[20] Eventually, the residents of Kentucky County petitioned for a separation from Virginia. Ten constitutional conventions were held in the Constitution Square Courthouse in Danville between 1784 and 1792. In 1790, Kentucky's delegates accepted Virginia's terms of separation, and a state constitution was drafted at the final convention in April 1792. On June 1, 1792, Kentucky became the fifteenth state to be admitted to the union. Isaac Shelby, a military veteran from Virginia, was elected the first Governor of the Commonwealth of Kentucky.[21]

19th century

Central Kentucky, the Bluegrass region, was the center of the greatest slaveholding, as planters cultivated tobacco and hemp, and also were noted for their quality livestock. During the 19th century, Kentucky slaveholders began to sell surplus slaves to the Deep South, with Louisville becoming a major slave market and departure port for slaves being transported downriver.

Designed by the Washington Monument's architect Robert Mills in 1845, the U.S. Marine Hospital in Louisville is considered the best remaining antebellum hospital in the United States.

It was one of the border states during the American Civil War.[22] Although frequently described as never having seceded, representatives from several counties met at Russellville calling themselves the "Convention of the People of Kentucky" and passed an Ordinance of Secession on November 20, 1861.[23] They established a Confederate government of Kentucky with its capital in Bowling Green.[24]

Though Kentucky was represented by the central star on the Confederate battle flag,[25] the Russellville Convention did not represent the majority of residents. Kentucky officially remained "neutral" throughout the war due to Union sympathies of many of the Commonwealth's citizens. In a revival of the "Lost Cause" that has exceeded the support it gained during the war, some contemporary people observe Confederate Memorial Day on Confederate President Jefferson Davis' birthday, June 3 and participate in Confederate re-enactments.[26] [27]

20th century

The Black Patch Tobacco Wars, a vigilante action, occurred in the area in the early 20th century. As result of the tobacco industry monopoly, tobacco farmers in the area were forced to sell their tobacco at low prices. Many local farmers and activists united to refuse to sell tobacco to the tobacco industry.

A vigilante wing, the "Night Riders", terrorized farmers who sold their tobacco at the low prices demanded by the tobacco corporations. They burned several tobacco warehouses, notably in Hopkinsville and Princeton. In the later period of their operation, they were known to physically assault farmers who broke the boycott. The Governor declared martial law and deployed the Kentucky Militia to end the Black Patch Tobacco Wars.

Statue of William Goebel in Frankfort

On January 30, 1900, Governor William Goebel, flanked by two bodyguards, was mortally wounded by an assassin while walking to the State Capitol in downtown Frankfort. Goebel was contesting the Kentucky gubernatorial election of 1899, which William S. Taylor was initially believed to have won. For several months, J. C. W. Beckham, Goebel's running mate, and Taylor fought over who was the legal governor, until the Supreme Court of the United States ruled in May in favor of Beckham. After fleeing to Indiana, Taylor was indicted as a co-conspirator in Goebel's assassination. Goebel is the only governor of a U.S. state to have been assassinated while in office.[28]

Law and government

Kentucky is one of four U.S. states to officially use the term *commonwealth*. The term was used for Kentucky as it had also been used by Virginia, from which Kentucky was created. The term has no particular significance in its meaning and was chosen to emphasize the distinction from the status of royal colonies as a place governed for the general welfare of the populace.[29]

The commonwealth term was used in citizen petitions submitted between 1786 and 1792 for the creation of the state, and in the Kentucky Constitution adopted in 1850. It was also used in the title of a history of the state that was published in 1834 and was used in various places within that book in references to Virginia and Kentucky.[30] The other two states officially called "commonwealths" are Massachusetts and Pennsylvania.

Kentucky is one of only five states that elects its state officials in odd-numbered years (the others being Louisiana, Mississippi, New Jersey, and Virginia). Kentucky holds elections for these offices every 4 years in the years preceding Presidential election years. Thus, Kentucky held gubernatorial elections in 2003, 2007, and 2011.

Executive branch

The governor's mansion in Frankfort

The executive branch is headed by the governor who serves as both head of state and head of government. The lieutenant governor may or may not have executive authority depending on whether the person is a member of the Governor's cabinet. Under the current Kentucky Constitution, the lieutenant governor assumes the duties of the governor only if the governor is incapacitated. (Prior to 1992, the lieutenant governor assumed power any time the governor was out of the state.) The governor and lieutenant governor usually run on a single ticket (also per a 1992 constitutional amendment), and are elected to four-year terms. Currently, the governor and lieutenant governor are Democrats Steve Beshear and Jerry Abramson.

Other elected constitutional offices include the Secretary of State, Attorney General, Auditor of Public Accounts, State Treasurer and Commissioner of Agriculture. Currently, Democrat Alison Lundergan Grimes serves as the Secretary of State. The commonwealth's chief prosecutor, law enforcement officer, and law officer is the Attorney General, currently Democrat Jack Conway. The Auditor of Public Accounts is Democrat Crit Luallen. Democrat Todd Hollenbach is the current Treasurer. Republican James Comer is the current Commissioner of Agriculture.

Legislative branch

The Kentucky State Capitol building in Frankfort

Kentucky's legislative branch consists of a bicameral body known as the Kentucky General Assembly.

The Senate is considered the upper house. It has 38 members, and is led by the President of the Senate, currently Republican Robert Stivers. Previous holder of the office, Republican David L. Williams, resigned from the Senate to accept a circuit court judgeship.

The House of Representatives has 100 members, and is led by the Speaker of the House, currently Democrat Greg Stumbo.

Judicial branch

The judicial branch of Kentucky is called the Kentucky Court of Justice [31] and comprises courts of limited jurisdiction called District Courts; courts of general jurisdiction called Circuit Courts; specialty courts such as Drug Court, Family Court [32]; an intermediate appellate court, the Kentucky Court of Appeals; and a court of last resort, the Kentucky Supreme Court.

The Kentucky Court of Justice is headed by the Chief Justice of the Commonwealth.

Unlike federal judges, who are usually appointed, justices serving on Kentucky state courts are chosen by the state's populace in non-partisan elections.

Federal representation

Kentucky's two U.S. Senators are Senate Minority Leader Mitch McConnell and Rand Paul, both Republicans. The state is divided into six Congressional Districts, represented by Republicans Ed Whitfield (1st), Brett Guthrie (2nd), Thomas Massie (4th), Hal Rogers (5th) and Andy Barr (6th) and Democrat John Yarmuth (3rd).

A map showing Kentucky's six congressional districts.

Judicially, Kentucky is split into two Federal court districts: the Kentucky Eastern District and the Kentucky Western District. Appeals are heard in the Sixth Circuit Court of Appeals based in Cincinnati, Ohio.

Law

Kentucky's body of laws, known as the Kentucky Revised Statutes (KRS), were enacted in 1942 to better organize and clarify the whole of Kentucky law.[33] The statutes are enforced by local police, sheriffs and deputy sheriffs, and constables and deputy constables. Unless they have completed a police academy elsewhere, these officers are required to complete training at the Kentucky Department of Criminal Justice Training Center on the campus of Eastern Kentucky University.[34] Additionally, in 1948, the Kentucky General Assembly established the Kentucky State Police, making it the 38th state to create a force whose jurisdiction extends throughout the given state.[35]

Kentucky is one of the 36 states in the United States that sanctions the death penalty for certain heinous murders. Those convicted of capital crimes after March 31, 1998 are always executed by lethal injection; those convicted on or before this date may opt for the electric chair.[36] Only three people have been executed in Kentucky since the U.S. Supreme Court re-instituted the practice in 1976. The most notable execution in Kentucky, however, was that of Rainey Bethea on August 14, 1936. Bethea was publicly hanged in Owensboro for the rape and murder of Lischia Edwards.[37] Irregularities with the execution led to this becoming the last public execution in the United States.[38]

Kentucky has been on the front lines of the debate over displaying the Ten Commandments on public property. In the 2005 case of *McCreary County v. ACLU of Kentucky*, the U.S. Supreme Court upheld the decision of the Sixth Circuit Court of Appeals that a display of the Ten Commandments in the Whitley City courthouse of McCreary County was unconstitutional.[39] Later that year, Judge Richard Fred Suhrheinrich, writing for the Sixth Circuit Court of Appeals in the case of *ACLU of Kentucky v. Mercer County*, wrote that a display including the Mayflower Compact, the Declaration of Independence, the Ten Commandments, the Magna Carta, *The Star-Spangled Banner*, and the national motto could be erected in the Mercer County courthouse.[40] Still, a 2008 study found that Kentucky's Supreme Court to be the least influential high court in the nation with its decisions rarely being followed by other states.[41]

Kentucky has also been known to have unusually high political candidacy age laws, especially compared to surrounding states. The origin of this is unknown, but it has been suggested it has to do with the commonwealth tradition.

Politics

Presidential elections results[42]

Year	Republican	Democrat
2012	**60.49%** *1,087,190*	37.80% *679,370*
2008	**57.37%** *1,048,462*	41.15% *751,985*
2004	**59.55%** *1,069,439*	39.69% *712,733*
2000	**56.50%** *872,492*	41.37% *638,898*
1996	44.88% *623,283*	**45.84%** *636,614*
1992	41.34% *617,178*	**44.55%** *665,104*
1988	**55.52%** *734,281*	43.88% *580,368*
1984	**60.04%** *822,782*	39.37% *539,589*
1980	**49.07%** *635,274*	47.61% *616,417*
1976	45.57% *531,852*	**52.75%** *615,717*
1972	**63.37%** *676,446*	34.77% *371,159*
1968	**43.79%** *462,411*	37.65% *397,541*
1964	35.65% *372,977*	**64.01%** *669,659*
1960	**53.59%** *602,607*	46.41% *521,855*

Where politics are concerned, Kentucky historically has been very hard-fought and leaned slightly toward the Democratic Party, although it was never included among the "Solid South". In 2006, 57.05% of the state's voters were officially registered as Democrats, 36.55% registered Republican, and 6.39% registered with some other political party.[43] Despite this, the state often supports Republican candidates for federal offices.

From 1964 through 2004, Kentucky voted for the eventual winner of the election for President of the United States. In the 2008 election, however, the state lost its bellwether status when John McCain, who won Kentucky, lost the national popular and electoral vote to Barack Obama (McCain carried Kentucky 57 to 41%).

The Commonwealth supported the previous three Democratic candidates elected to the White House, all elected from Southern states: Lyndon B. Johnson (Texas) in 1964, Jimmy Carter (Georgia) in 1976, and Bill Clinton (Arkansas) in 1992 and 1996. In presidential elections, the state has become a Republican stronghold, supporting that party's presidential candidates by double-digit margins in 2000, 2004, 2008 and 2012.

Voter Registration and Party Enrollment as of June 26, 2010[44]

Party		Number of Voters	Percentage
	Democratic	1,619,391	57%
	Republican	1,052,902	37%
	Other	189,499	6%
Total		**2,861,792**	**100%**

Demographics

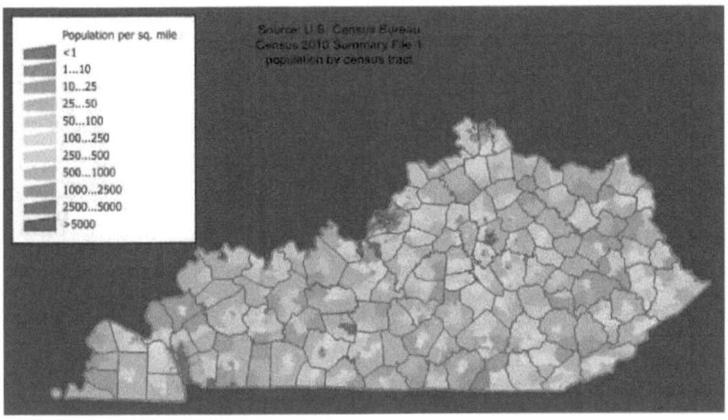

Kentucky Population Density Map

Historical populations			
Census	**Pop.**		**%±**
1790	73677		—
1800	220955		199.9%
1810	406511		84.0%
1820	564317		38.8%
1830	687917		21.9%
1840	779828		13.4%
1850	982405		26.0%
1860	1155684		17.6%
1870	1321011		14.3%
1880	1648690		24.8%
1890	1858635		12.7%
1900	2147174		15.5%
1910	2289905		6.6%
1920	2416630		5.5%
1930	2614589		8.2%
1940	2845627		8.8%
1950	2944806		3.5%
1960	3038156		3.2%
1970	3218706		5.9%
1980	3660777		13.7%
1990	3685296		0.7%
2000	4041769		9.7%
2010	4339367		7.4%
Est. 2012	4380415		0.9%

Source: 1790–2000[45]
1910–2010[46]

The United States Census Bureau estimates that the population of Kentucky was 4,380,415 on July 1, 2012, a 0.9% increase since the 2010 United States Census.[]

As of July 1, 2012, Kentucky has an estimated population of 4,380,415, which is an increase of 11,059, or 0.3%, from the prior year and an increase of 41,058, or 0.9%, since the year 2010. This includes a natural increase since the last census of 77,156 people (that is 287,222 births minus 210,066 deaths) and an increase due to net migration of 59,604 people into the state. Immigration from outside the United States resulted in a net increase of 27,435 people, and migration within the country produced a net increase of 32,169 people. As of 2004, Kentucky's population included about 95,000 foreign-born persons (2.3%). The population density of the state is 101.7 people per square mile.[47]

Kentucky's total population has grown during every decade since records began. However, during most decades of the 20th century there was also net out-migration from Kentucky. Since 1900, rural Kentucky counties have experienced a net loss of over 1 million people from migration, while urban areas have experienced a slight net gain.[48]

The center of population of Kentucky is located in Washington County, in the city of Willisburg.[49]

Race and ancestry

The largest ancestries in the commonwealth are: English (30.6%), German (12.7%), Irish (10.5%), and African (7.8%).[50] [51] In the state's most urban counties of Jefferson, Oldham, Fayette, Boone, Kenton, and Campbell, German is the largest reported ancestry. Americans of Scots-Irish and English stock are present throughout the entire state. Many claim Irish ancestry because of the term "Scots-Irish". Southeastern Kentucky was populated by a large group of multiracial settlers, sometimes called Melungeons, in the early 19th century.

Groups such as the Ridgetop Shawnee in the early 21st century organized as a non-profit to increase awareness of Native American descent in Kentucky. In the 2000 census, there were 20,000 people in the state who identified as Native American (0.49%). In June 2011, Jerry “2 Feather” Thornton, a Cherokee, led a team in the Voyage of Native American Awareness 2011 canoe journey, to begin on the Green River in Rochester, Kentucky and travel through to the Ohio River at Henderson, Kentucky.[52]

African Americans, who made up one-fourth of Kentucky's population prior to the Civil War, primarily in the Bluegrass region, declined in number during the 20th century, as many moved to the industrial North in the Great Migration. Today, 44.2% of Kentucky's African-American population is in Jefferson County and 52% are in the Louisville Metro Area; 20% of the county's population is African American. Other areas with high concentrations, beside Christian and Fulton counties, are the city of Paducah, the Bluegrass, and the city of Lexington. Some mining communities in far Southeastern Kentucky have populations that are between five and 10 percent African American.

Race/Ethnicity (2010)	
White, non-Hispanic	86.3%
Black or African American	7.8%
Hispanic or Latino (of any race)	3.1%
Asian	1.1%
American Indian and Alaska Native	0.2%
Pacific Islander	0.1%

Religion

In 2000, The Association of Religion Data Archives reported[] that of Kentucky's 4,041,769 residents:

- 47% were not affiliated with any church
- 34% were members of Evangelical Protestant churches
  - Southern Baptist Convention (979,994 members, 24%)
  - Christian churches and churches of Christ (106,638 members, 3%)
  - Churches of Christ (58,602 members, 1%)
- 10% were Roman Catholics
- 9% belonged to mainline Protestant churches
  - United Methodist Church (208,720 members, 5%)
  - Christian Church (Disciples of Christ) (67,611 members, 2%)
- 0.05% were Orthodox Christians
- 1% were affiliated with other theologies.

Lexington Theological Seminary (then College of the Bible), 1904.

Today Kentucky is home to several seminaries. Southern Baptist Theological Seminary in Louisville is the principal seminary for the Southern Baptist Convention. Louisville is also the home of the Louisville Presbyterian Theological Seminary, an institution of the Presbyterian Church (U.S.A.) [PC(USA)]. Lexington has two seminaries, Lexington Theological Seminary (affiliated with the Disciples of Christ), and the Baptist Seminary of Kentucky. Asbury Theological Seminary, a multi-denominational seminary in the Methodist tradition, is located in nearby Wilmore. In addition to seminaries, there are several colleges affiliated with denominations. Transylvania in Lexington is affiliated with the Disciples of Christ. The University of Pikeville in Pikeville is affiliated with PC(USA).

In Louisville, Bellarmine and Spalding are affiliated with the Roman Catholic Church. In Owensboro, Kentucky, Kentucky Wesleyan College is associated with the United Methodist Church and Brescia University is associated with the Roman Catholic Church. Wilmore is home to Asbury University (a separate institution from the seminary), which is associated with the Christian College Consortium. The University of the Cumberlands, located in Williamsburg, Campbellsville University in Campbellsville, Georgetown College in Georgetown and Mid-Continent University in Mayfield all have connections with the Southern Baptist Convention.

Louisville is home to the headquarters of the PC(USA) and their printing press. Louisville is also home to Muslim,[53] Jewish, and Hindu communities.

There is also a growing Hindu population in Hopkinsville, KY.

Economy

Early in its history Kentucky gained recognition for its excellent farming conditions. It was the site of the first commercial winery in the United States (started in present day Jessamine County in 1799) and due to the high calcium content of the soil in the Bluegrass region quickly became a major horse breeding (and later racing) area. Today Kentucky ranks 5th nationally in goat farming, 8th in beef cattle production,[54] and 14th in corn production.[55]

The best selling car in the United States, the Toyota Camry, is manufactured in Georgetown, Kentucky.

Today Kentucky's economy has expanded to importance in non agricultural terms as well, especially in auto manufacturing, energy fuel production, and medical facilities. As of 2010 24% of electricity produced in the USA depended on either enriched uranium rods coming from the Paducah Gaseous Diffusion Plant (the only domestic site of low grade uranium enrichment), or from the 107,336 tons of coal extracted from the state's two coal fields (which combined produce 4% percent of the electricity in the United States).[56]

The best selling truck in the United States, the Ford F-Series, is manufactured in Louisville, Kentucky.

Kentucky ranks 4th among U.S. states in the number of automobiles and trucks assembled.[57] The Chevrolet Corvette, Cadillac XLR (2004–2009), Ford Escape, Ford Super Duty trucks, Ford Expedition, Lincoln Navigator, Toyota Camry, Toyota Avalon, Toyota Solara, and Toyota Venza are assembled in Kentucky.

Kentucky exports reached a record $22.1 billion in 2012, with products and services going to 199 countries.[58]

According to the Kentucky Cabinet for Economic Development, the primary state agency in Kentucky responsible for creating new jobs and new investment in the state, new business investment in Kentucky in 2012 totaled nearly $2.7 billion, with the creation of more than 14,000 new jobs. One such investment was L'Oreal in Northern Kentucky,which added 200 jobs on top of the 280 already in existing facilities in Florence and Walton.[59]

In May 2010, the Army Human Resource Center of Excellence, the largest office building in the state at nearly 900,000-square-feet, opened at Fort Knox. The new complex employs nearly 4,300 soldiers and civilians.[60]

The total gross state product for 2010 was $163.3 billion, 28th in the nation.[61] Its per-capita personal income was US$28,513, 43rd in the nation.[62] An organization called the Institute for Truth in Accounting estimated that the state government's debts exceeded its available assets by US$26,300 per taxpayer as of 2011, ranking the state as having the 5th highest such debt burden in the nation.[]

As of July 2012, the state's unemployment rate is 8.3%.[63]

Taxation

There are 6 income tax brackets, ranging from 2% to 6% of personal income.[64] The sales tax rate in Kentucky is 6%.[65] Kentucky has a broadly based classified property tax system. All classes of property, unless exempted by the Constitution, are taxed by the state, although at widely varying rates.[66]

Many of these classes are exempted from taxation by local government. Of the classes that are subject to local taxation, three have special rates set by the General Assembly, one by the Kentucky Supreme Court and the remaining classes are subject to the full local rate, which includes the tax rate set by the local taxing bodies plus all voted levies. Real property is assessed on 100% of the fair market value and property taxes are due by December 31. Once the primary source of state and local government revenue, property taxes now account for only about 6% of the

Kentucky's annual General Fund revenues.[67]

Until January 1, 2006, Kentucky imposed a tax on intangible personal property held by a taxpayer on January 1 of each year. The Kentucky intangible tax was repealed under House Bill 272.[68] Intangible property consisted of any property or investment which represents evidence of value or the right to value. Some types of intangible property included: bonds, notes, retail repurchase agreements, accounts receivable, trusts, enforceable contracts sale of real estate (land contracts), money in hand, money in safe deposit boxes, annuities, interests in estates, loans to stockholders, and commercial paper.

Government-promoted slogans

In December 2002, the Kentucky governor Paul Patton unveiled the state slogan "It's that friendly",[] in hope of drawing more people into the state based on the idea of southern hospitality. This campaign was neither a failure nor a success. Though it was meant to embrace southern values, many Kentuckians rejected the slogan as cheesy and ineffective.[] It was quickly seen that the slogan did not encourage tourism as much as initially hoped for. So government decided to create a different slogan to embrace Kentucky as a whole while also encouraging more people to visit the Bluegrass.[69]

To boost Kentucky's image, give it a consistent reach, and help it "stand out from the crowd", former Governor Ernie Fletcher launched a comprehensive branding campaign at some later point in time with the hope of making the state's $12–14 million advertising budget more effective. The resulting "Unbridled Spirit" brand was the result of a $500,000 contract with New West, a Kentucky-based public relations advertising and marketing firm to develop a viable brand and tag line.[70] The Fletcher administration aggressively marketed the brand in both the public and private sectors. Since that time, the "Welcome to Kentucky" signs at border areas have an "Unbridled Spirit" symbol on them.

Transportation

Roads

Kentucky is served by five major interstate highways (I-75, I-71, I-64, I-65, I-24), nine parkways, and three bypasses and spurs. The parkways were originally toll roads, but on November 22, 2006, Governor Ernie Fletcher ended the toll charges on the William H. Natcher Parkway and the Audubon Parkway, the last two parkways in Kentucky to charge tolls for access.[71] The related toll booths have been demolished.[72]

At 464 miles (km) long, Kentucky Route 80 is the longest route in Kentucky, pictured here west of Somerset.

Ending the tolls some seven months ahead of schedule was generally agreed to have been a positive economic development for transportation in Kentucky. In June 2007, a law went into effect raising the speed limit on rural portions of Kentucky Interstates from 65 to 70 miles per hour (**unknown operator: u'['** to **unknown operator: u'['** km/h).[73]

Road tunnels include the interstate Cumberland Gap Tunnel and the rural Nada Tunnel.

Greyhound provides bus service to most major cities in the state.

Rails

Amtrak, the national passenger rail system, provides service to Ashland, South Portsmouth, Maysville and Fulton. The *Cardinal* (trains 50 and 51) is the line that offers Amtrak service to Ashland, South Shore, Maysville and South Portsmouth. The *City of New Orleans* (trains 58 and 59) serve Fulton. The Northern Kentucky area is served by the *Cardinal* at the Cincinnati Museum Center at Union Terminal. The Museum Center is just across the Ohio River in Cincinnati.

High Bridge over the Kentucky River was the tallest rail bridge in the world when it was completed in 1877.

As of 2004, there were approximately 2640 miles (km) of railways in Kentucky, with about 65% of those being operated by CSX Transportation. Coal was by far the most common cargo, accounting for 76% of cargo loaded and 61% of cargo delivered.[74]

Bardstown features a tourist attraction known as *My Old Kentucky Dinner Train*. Run along a 20-mile (km) stretch of rail purchased from CSX in 1987, guests are served a four-course meal as they make a two-and-a-half hour round-trip between Bardstown and Limestone Springs.[75] The Kentucky Railway Museum is located in nearby New Haven.[76]

Other areas in Kentucky are reclaiming old railways in rail trail projects. One such project is Louisville's Big Four Bridge. If completed, the Big Four Bridge rail trail will contain the second longest pedestrian-only bridge in the world.[77] The longest pedestrian-only bridge is also found in Kentucky—the Newport Southbank Bridge, popularly known as the "Purple People Bridge", connecting Newport to Cincinnati, Ohio.[78]

Air

Kentucky's primary airports include Louisville International Airport (Standiford Field (SDF)) of Louisville, Cincinnati/Northern Kentucky International Airport (CVG) of Covington and Northern Kentucky, and Blue Grass Airport (LEX) in Lexington. Louisville International Airport is home to UPS's Worldport, its international air-sorting hub.[79] There are also a number of regional airports scattered across the state.

On August 27, 2006, Kentucky's Blue Grass Airport in Lexington was the site of a crash that killed 47 passengers and 2 crew members aboard a Bombardier Canadair Regional Jet designated Comair Flight 191, or Delta Air Lines Flight 5191, sometimes mistakenly identified by the press as Comair Flight 5191.[80] The lone survivor was the flight's first officer, James Polehinke, who doctors determined to be brain damaged and unable to recall the crash at all.[81]

Water

A barge hauling coal in the Louisville and Portland Canal, the only manmade section of the Ohio River.

As the state is bounded by two of the largest rivers in North America, water transportation has historically played a major role in Kentucky's economy. Louisville was a major port for steamships in the nineteenth century. Today, most barge traffic on Kentucky waterways consists of coal that is shipped from both the Eastern and Western Coalfields, about half of which is used locally to power many power plants located directly off the Ohio River, with the rest being exported to other countries, most notably Japan.

Many of the largest ports in the United States are located in or adjacent to Kentucky, including:

- Huntington/Tri-State (includes Ashland, KY), largest inland port and 7th largest overall
- Cincinnati-Northern Kentucky, 5th largest inland port and 43rd overall
- Louisville-Southern Indiana, 7th largest inland port and 55th overall

As a state, Kentucky ranks 10th overall in port tonnage.[82] [83]

The only natural obstacle along the entire length of the Ohio River is the Falls of the Ohio, located just west of Downtown Louisville.

Subdivisions and settlements

Counties

Map of Kentucky counties

Kentucky is subdivided into 120 counties, the largest being Pike County at 787.6 square miles (km^2), and the most populous being Jefferson County (which coincides with the Louisville Metro governmental area) with 741,096 residents as of 2010.[85]

County government, under the Kentucky Constitution of 1891, is vested in the County Judge/Executive, (formerly called the County Judge) who serves as the executive head of the county, and a legislature called a Fiscal Court. Despite the unusual name, the Fiscal Court no longer has judicial functions.

Consolidated city-county governments

Kentucky's two most populous counties, Jefferson and Fayette, have their governments consolidated with the governments of their largest cities. *Louisville-Jefferson County Government* (Louisville Metro) and *Lexington-Fayette Urban County Government* (Lexington Metro) are unique in that their city councils and county Fiscal Court structures have been merged into a single entity with a single chief executive, the Metro Mayor and Urban County Mayor, respectively. Although the counties still exist as subdivisions of the state, in reference the names Louisville and Lexington are used to refer to the entire area coextensive with the former cities and counties. Somewhat incongruously, when entering Lexington-Fayette the highway signs read "Fayette County" while most signs leading into Louisville-Jefferson simply read "Welcome to Louisville Metro."

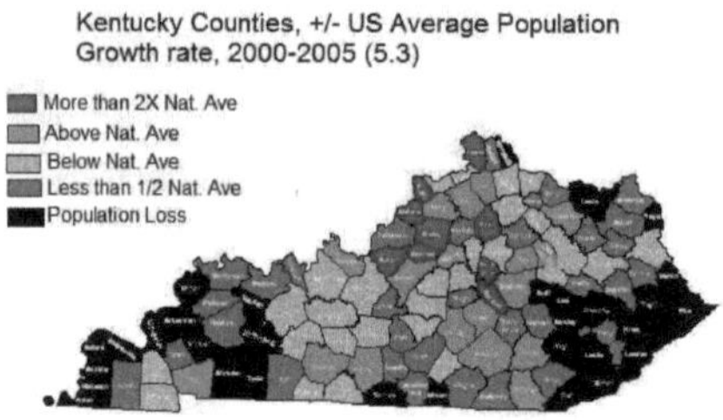

Population growth [84] is centered along and between interstates I-65 and I-75.

Louisville is the state's largest city with a metro population of 1.3 million

Lexington is the state's second largest city with a metro population of around 500,000

Covington is the largest city in Northern Kentucky

Cities and towns

Rank	City	2010 Pop	2000 Pop	Δ Current Pop
1	Louisville	765,515	551,299	214,216
2	Lexington	295,803	260,512	35,291
3	Bowling Green	58,067	49,296	8,771
4	Owensboro	57,265	54,067	3,198
5	Covington	40,640	43,370	-2,730
6	Hopkinsville	31,577	30,089	1,488
7	Richmond	31,364	27,152	4,212
8	Florence	29,951	23,551	6,400
9	Georgetown	29,098	18,080	11,018
10	Henderson	28,757	27,373	1,384
11	Elizabethtown	28,531	22,542	5,989
12	Nicholasville	28,015	19,680	8,335
13	Jeffersontown	26,595	26,442	153
14	Frankfort	25,527	27,741	-2,214
15	Paducah	25,024	26,442	-1,418

The Metro Louisville government area has a 2010 population of 741,096. Under United States Census Bureau methodology, the population of Louisville was 566,503. The latter figure is the population of the so-called "balance"—the parts of Jefferson County that were either unincorporated or within the City of Louisville before the formation of the merged government in 2003. In 2010, the Louisville Combined Statistical Area (CSA) has a population of 1,451,564; including 1,061,031 in Kentucky, which is nearly one-fourth of the state's population. Since 2000, over one-third of the state's population growth has occurred in the Louisville CSA. In addition, the top 28 wealthiest places in Kentucky are in Jefferson County and seven of the 15 wealthiest counties in the state are located in the Louisville CSA.[86]

The second largest city is Lexington with a 2010 census population of 295,803 and its CSA, which includes the Frankfort and Richmond statistical areas, having a population of 687,173. The Northern Kentucky area (the seven Kentucky counties in the Cincinnati MSA) had a population of 425,483 in 2010. The metropolitan areas of

Louisville, Lexington, and Northern Kentucky have a combined population of 2,173,687 as of 2010, which is 50.1% of the state's total population.

The two other fast growing urban areas in Kentucky are the Bowling Green area and the "Tri Cities Region" of southeastern Kentucky, comprising Somerset, London and Corbin.

Although only one town in the "Tri Cities", namely Somerset, currently has more than 12,000 people, the area has been experiencing heightened population and job growth since the 1990s. Growth has been especially rapid in Laurel County, which outgrew areas such as Scott and Jessamine counties around Lexington or Shelby and Nelson Counties around Louisville. London significantly grew in population in the 2000s, from 5,692 in 2000 to 7,993 in 2010. London also landed a Wal-Mart distribution center in 1997, bringing thousands of jobs to the community.

In northeast Kentucky, the greater Ashland area is an important transportation, manufacturing, and medical center. Iron and petroleum production, as well as the transport of coal by rail and barge, have been historical pillars of the region's economy. Due to a decline in the area's industrial base, Ashland has seen a sizable reduction in its population since 1990. The population of the area has since stabilized, however, with the medical service industry taking a greater role in the local economy. The Ashland area, including the counties of Boyd and Greenup, are part of the Huntington-Ashland, WV-KY-OH, Metropolitan Statistical Area (MSA). As of the 2000 census, the MSA had a population of 288,649. More than 21,000 of those people (as of 2010) reside within the city limits of Ashland.

The largest county in Kentucky by area is Pike, which contains Pikeville and suburb Coal Run Village . The county and surrounding area is the most populated region in the state that is not part of a Micropolitan Statistical Area or a Metropolitan Statistical Area containing nearly 200,000 people in five counties: Floyd County, Martin County, Letcher County, and neighboring Mingo County, West Virginia. Pike County contains slightly over 68,000 people.

Only three U.S. states have capitals with smaller populations than Kentucky's Frankfort (pop. 25,527), those being Augusta, Maine (pop. 18,560), Pierre, South Dakota (pop. 13,876), and Montpelier, Vermont (pop. 8,035).

Education

Kentucky maintains eight public four-year universities. There are two general tiers: major research institutions (the University of Kentucky and the University of Louisville) and regional universities, which encompasses the remaining 6 schools. The regional schools have specific target counties that many of their programs are targeted towards (such as Forestry at Eastern Kentucky University or Cave Management at Western Kentucky University), however most of their curriculum varies little from any other public university.

The University of Kentucky is Kentucky's flagship university.

UK and UofL have the highest academic rankings and admissions standards although the regional schools aren't without their national recognized departments – examples being Western Kentucky University's nationally ranked Journalism Department or Morehead State University offering one of the nation's only Space Science degrees. UK is the flagship and land grant of the system and has agriculture extension services in every county. The two research schools split duties related to the medical field, UK handles all medical outreach programs in the eastern half of the state while UofL does all medical outreach in the state's western half.

The state's sixteen public two-year colleges have been governed by the Kentucky Community and Technical College System since the passage of the Postsecondary Education Improvement Act of 1997, commonly referred to as House Bill 1.[87] Prior to the passage of House Bill 1, most of these colleges were under the control of the University of Kentucky.

The University of Louisville is Kentucky's urban research university.

Transylvania University, located in Lexington, is the oldest university west of the Allegheny Mountains, founded in 1780. Transylvania is a liberal arts university, consistently ranked in the top tier in the country.

Berea College, located at the extreme southern edge of the Bluegrass below the Cumberland Plateau, was the first coeducational college in the South to admit both black and white students, doing so from its very establishment in 1855.[88] This policy was successfully challenged in the United States Supreme Court in the case of *Berea College v. Kentucky* in 1908.[89] This decision effectively segregated Berea until the landmark *Brown v. Board of Education* in 1954.

There are 174 school districts and 1,233 public schools in Kentucky.[90] For the 2010 to 2011 school year, there were approximately 647,827 students enrolled in public school.[91]

Kentucky has been the site of much educational reform over the past two decades. In 1989, the Kentucky Supreme Court ruled that the state's education system was unconstitutional.[92] The response of the General Assembly was passage of the Kentucky Education Reform Act (KERA) the following year. Years later, Kentucky has shown progress, but most agree that further reform is needed.[93]

Culture

Although Kentucky's culture is generally considered to be Southern, it is unique in that it is also influenced by the Midwest and Southern Appalachia in certain areas of the state. The state is known for bourbon and whiskey distilling, tobacco, horse racing, and college basketball. Kentucky is more similar to the Upland South in terms of ancestry which is predominantly American.[94]

The Buffalo Trace Distillery

Nevertheless, during the 19th century, Kentucky did receive a substantial number of German immigrants, who settled mostly in the Midwest, along the Ohio River primarily in Louisville, Covington and Newport.[95] Only Maryland, Delaware and West Virginia have higher German ancestry percentages than Kentucky among Census-defined Southern states, although Kentucky's percentage is closer to Arkansas and Virginia's than the previously named state's percentages. Scottish Americans, English Americans and Scotch-Irish Americans have heavily influenced Kentucky culture, and are present in every part of the state.[96]

Kentucky was a slave state, and blacks once comprised over one-quarter of its population. However, it lacked the cotton plantation system and never had the same high percentage of African Americans as most other slave states. With less than 8% of its current population being black, Kentucky is rarely included in modern-day definitions of the Black Belt, despite a relatively significant rural African American population in the Central and Western areas of the state.[97] [98] [99]

Kentucky adopted the Jim Crow system of racial segregation in most public spheres after the Civil War, but the state never disenfranchised African American citizens to the level of the Deep South states, and it peacefully integrated its schools after the 1954 *Brown v. Board of Education* verdict, later adopting the first state civil rights act in the South in 1966.[100]

Old Louisville is the largest Victorian Historic neighborhood in the United States.

The biggest day in American horse racing, the Kentucky Derby, is preceded by the two-week Derby Festival[101] in Louisville. Louisville also plays host to the Kentucky State Fair,[102] the Kentucky Shakespeare Festival,[103] and Southern gospel's annual highlight, the National Quartet Convention.[104] Bowling Green, the state's third-largest city and home to the only assembly plant in the world that manufactures the Chevrolet Corvette,[105] opened the National Corvette Museum in 1994.[106] The fourth-largest city, Owensboro, gives credence to its nickname of "Barbecue Capital of the World" by hosting the annual International Bar-B-Q Festival.[107]

Old Louisville, the largest historic preservation district in the United States featuring Victorian architecture and the third largest overall,[108] hosts the St. James Court Art Show, the largest outdoor art show in the United States.[109] The neighborhood was also home to the Southern Exposition (1883–1887), which featured the first public display of Thomas Edison's light bulb,[110] and was the setting of Alice Hegan Rice's novel, *Mrs. Wiggs of the Cabbage Patch* and Fontaine Fox's comic strip, the "Toonerville Trolley.[111]

Hodgenville, the birthplace of Abraham Lincoln, hosts the annual Lincoln Days Celebration, and also hosted the kick-off for the National Abraham Lincoln Bicentennial Celebration in February 2008. Bardstown celebrates its heritage as a major bourbon-producing region with the Kentucky Bourbon Festival.[112] Glasgow mimics Glasgow, Scotland by hosting the Glasgow Highland Games, its own version of the Highland Games,[113] and Sturgis hosts "Little Sturgis", a mini version of Sturgis, South Dakota's annual Sturgis Motorcycle Rally.[114]

The residents of tiny Benton pay tribute to their favorite tuber, the sweet potato, by hosting Tater Day.[115] Residents of Clarkson in Grayson County celebrate their city's ties to the honey industry by celebrating the Clarkson Honeyfest.[116] The Clarkson Honeyfest is held the last Thursday, Friday and Saturday in September, and is the "Official State Honey Festival of Kentucky."

Music

The breadth of music in Kentucky is indeed wide, stretching from the Purchase to the eastern mountains.

Renfro Valley, Kentucky is home to Renfro Valley Entertainment Center and the Kentucky Music Hall of Fame and is known as "Kentucky's Country Music Capital," a designation given it by the Kentucky State Legislature in the late 1980s. The Renfro Valley Barn Dance was where Renfro Valley's musical heritage began, in 1939, and influential country music luminaries like Red Foley, Homer & Jethro, Lily May Ledford & the Original Coon Creek Girls, Martha Carson, and many others have performed as regular members of the shows there over the years. The Renfro Valley Gatherin' is today America's second oldest continually broadcast radio program of any kind. It is broadcast on local radio station WRVK and a syndicated network of nearly 200 other stations across the United States and Canada every week.

Contemporary Christian music star Steven Curtis Chapman is a Paducah native, and Rock and Roll Hall of Famers The Everly Brothers are closely connected with Muhlenberg County, where older brother Don was born. Merle Travis, Country & Western artist known for both his signature "Travis picking" guitar playing style, as well as his hit song "Sixteen Tons", was also born in Muhlenberg County. Kentucky was also home to Mildred and Patty Hill, the Louisville sisters credited with composing the tune to the ditty Happy Birthday to You in 1893; Loretta Lynn (Johnson County), Brian Littrell and Kevin Richardson of the Backstreet Boys, and Billy Ray Cyrus (Flatwoods).

The U.S. 23 Country Music Highway Museum in Paintsville provides background on the country music artists from Eastern Kentucky

However, its depth lies in its signature sound—Bluegrass music. Bill Monroe, "The Father of Bluegrass", was born in the small Ohio County town of Rosine, while Ricky Skaggs, Keith Whitley, David "Stringbean" Akeman, Louis Marshall "Grandpa" Jones, Sonny and Bobby Osborne, and Sam Bush (who has been compared to Monroe) all hail from Kentucky. The International Bluegrass Music Museum is located in Owensboro,[117] while the annual Festival of the Bluegrass is held in Lexington.[118]

Kentucky is also home to famed jazz musician and pioneer, Lionel Hampton (although this has been disputed in recent years).[119] Blues legend W.C. Handy and R&B singer Wilson Pickett also spent considerable time in Kentucky. The R&B group Midnight Star and Hip-Hop group Nappy Roots were both formed in Kentucky, as were country acts The Kentucky Headhunters, Montgomery Gentry and Halfway to Hazard, The Judds, as well as Dove Award-winning Christian groups Audio Adrenaline (rock) and Bride (metal). Heavy Rock band Black Stone Cherry hails from rural Edmonton, Indie rock band My Morning Jacket with lead singer and guitarist Jim James also originated out of Louisville, on the local independent music Scene. Rock bands Cage the Elephant, Sleeper Agent, and Morning Teleportation are also from Bowling Green. The bluegrass groups Driftwood and Kentucky Rain, along with Nick Lachey of the pop band 98 Degrees are also from Kentucky.

In eastern Kentucky, old-time music carries on the tradition of ancient ballads and reels developed in historical Appalachia.

Cuisine

The Hot Brown

Kentucky's cuisine is generally similar to traditional southern cooking, although in some areas of the state it can blend elements of both the South and Midwest.[120] [121] One original Kentucky dish is called the Hot Brown, a dish normally layered in this order: toasted bread, turkey, bacon, tomatoes and topped with mornay sauce. It was developed at the Brown Hotel in Louisville.[122] The Pendennis Club in Louisville is the birthplace of the *Old Fashioned* cocktail. Also, western Kentucky is known for its own regional style of barbecue.

Harland Sanders originated Kentucky Fried Chicken at his service station in North Corbin, though the first franchised KFC was located in South Salt Lake, Utah.[123]

Sports

Kentucky's Churchill Downs hosts the Kentucky Derby

Kentucky is the home of several sports teams such as Minor League Baseball's Triple-A Louisville Bats and Class A Lexington Legends and the Class A Bowling Green Hot Rods. They are also home to the Frontier Leagues Florence Freedom and several teams in the MCFL. The Lexington Horsemen and Louisville Fire of the now-defunct af2 had been interested in making a move up to the "major league" Arena Football League, but nothing has come of those plans.

The northern part of the state lies across the Ohio River from Cincinnati, which is home to a National Football League team, the Bengals, and a Major League Baseball team, the Reds. It is not uncommon for fans to park in the city of Newport and use the Newport Southbank Pedestrian Bridge, locally known as the "Purple People Bridge," to walk to these games in Cincinnati. Also, Georgetown College in Georgetown was the location for the Bengals' summer training camp, until it was announced in 2012 that the Bengals would no longer use the facilities.[124]

As in many states, especially those without major league professional sport teams, college athletics are very important. This is especially true of the state's three Division I Football Bowl Subdivision (FBS) programs, including the Kentucky Wildcats, the Western Kentucky Hilltoppers, and the Louisville Cardinals. The Wildcats, Hilltoppers, and Cardinals are among the most tradition-rich college men's basketball teams in the United States, combining for 11 National championships and 24 NCAA Final Fours; all three are high on the lists of total all-time wins, wins per season, and average wins per season.

The Kentucky Wildcats are particularly notable, leading all Division I programs in all-time wins, win percentage, NCAA tournament appearances, and being second only to UCLA in NCAA championships. Louisville has also stepped onto the football scene in recent years, including winning the 2007 Orange Bowl as well as the 2013 Allstate

Sugar Bowl. Western Kentucky, the 2002 national champion in Division I-AA football (now Football Championship Subdivision (FCS)), completed its transition to Division I FBS football in 2009.

Ohio Valley Wrestling in Louisville was the primary location for training and rehab for WWE professional wrestlers from 2000 until February 2008, when WWE ended its relationship with OVW and moved all of its contracted talent to Florida Championship Wrestling (now known as NXT Wrestling). In November 2011, OVW became the primary developmental territory for Total Nonstop Action Wrestling (TNA).

The NASCAR Sprint Cup Series has a race at the Kentucky Speedway in Sparta, Kentucky, which is within an hour driving distance from Cincinnati, Louisville and Lexington. The race is called the Quaker State 400. The NASCAR Nationwide Series and the Camping World Truck Series also race there.

State symbols

Insignia	Symbol	Binomial nomenclature	Year Adopted[125]
Official State Bird	Cardinal	*Cardinalis cardinalis*	1926
Official State Butterfly	Viceroy Butterfly	*Limenitis archippus*	1990
Official State Dance	Clogging		2006
Official State Beverage	Milk		2005
Official State Fish	Kentucky Spotted Bass	*Micropterus punctulatus*	2005
Official State Fossil	Brachiopod	undetermined	1986
Official State Flower	Goldenrod	*Soldiago gigantea*	1926
Official State Fruit	Blackberry	*Rubus allegheniensis*	2004
Official State Gemstone	Freshwater Pearl		1986
State Grass	Kentucky Bluegrass	*Poa pratensis*	Traditional
Official State Motto	*"United We Stand Divided We Fall"*		1942/1792
Official State Latin Motto	*"Deo gratiam habeamus"* ("Let us be grateful to God")		2002
Official State Horse	Thoroughbred	*Equus caballus*	1996
Official State Mineral	Coal		1998
Official State Outdoor Musical	"The Stephen Foster Story" (now called "Stephen Foster – The Musical")		2002
Official State Instrument	Appalachian Dulcimer		2001
State Nickname	"The Bluegrass State"		Traditional
Official State Rock	Kentucky Agate		2000
Official State Slogan			2004[126]
Official State Soil	Crider Soil Series		1990
Official State Tree	Tulip Poplar	*Liriodendron tulipifera*	1994
Official Wild Animal Game Species	Gray Squirrel	*Sciurus carolinensis*	1968
Official State Song	"My Old Kentucky Home" (revised version)		1986
Official State Silverware Pattern	*Old Kentucky Blue Grass: The Georgetown Pattern*		1996
Official State Music	Bluegrass music		2007[127]

Official State Automobile	Chevrolet Corvette		2010

Official state places and events

- State arboretum: Bernheim Arboretum and Research Forest
- State botanical garden: The Arboretum: State Botanical Garden of Kentucky
- State Science Center: Louisville Science Center
- State center for celebration of African American heritage: Kentucky Center for African American Heritage
- State honey festival: Clarkson Honeyfest[128]
- State amphitheater: Iroquois Amphitheater (Louisville)
- State tug-o-war championship: The Fordsville Tug-of-War Championship
- Covered Bridge Capital of Kentucky: Fleming County
- Official Covered Bridge of Kentucky: Switzer Covered Bridge (Franklin County)
- Official steam locomotive of Kentucky: "Old 152" (located in the Kentucky Railway Museum in New Haven)
- Official pipe band: Louisville Pipe Band
- State bourbon festival: Kentucky Bourbon Festival, Inc., in Bardstown, Kentucky
- Official State outdoor musical: The Stephen Foster Story at My Old Kentucky Home, Bardstown

Unless otherwise specified, all state symbol information is taken from Kentucky State Symbols [129].

Gallery

The world famous Louisville Slugger baseball bat is made in Kentucky. It holds the Guinness World Record for the largest bat.

Kentucky's 2001 commemorative quarter.

Thunder Over Louisville is the largest annual fireworks show in the world.

Kentucky's horse farms are world renowned.

The Daniel Boone National Forest.

The Ohio River forms the northern border of Kentucky.

Many Kentucky cities have historic areas near downtown, such as this example in Bowling Green.

See also

- Index of Kentucky-related articles
- Outline of Kentucky – organized list of topics about Kentucky

References

[8] http://www.enquirer.com/flood/floodphotos.html
[17] (http://cherokee.org/Issues/FraudulentGroups/31842/Information.aspx)
[27] Tony Hiss, *Confederates in the Attic*
[29] The Commonwealth of Kentucky (http://www.uky.edu/KentuckyAtlas/kentucky.html), Kentucky Atlas & Gazeteer, University of Kentucky website.
[30] Butler, Mann, *A History of the Commonwealth of Kentucky* (http://archive.org/details/historyofcommonw00butl), Wilcox, Dickerman & Co., 1834.
[31] http://courts.ky.gov/Pages/default.aspx
[32] http://courts.ky.gov/stateprograms/
[52] Vincent Schilling, "A Canoe Voyage to Increase Cultural Understanding" (http://indiancountrytodaymedianetwork.com/article/a-canoe-voyage-to-increase-cultural-understanding-24554), *Indian Country Today,* 19 May 2011, accessed 16 January 2012
[70] http://www.nytimes.com/2004/12/02/business/02addes.html?_r=0
[74] Also, Norfolk Southern's main north-south line runs through central and southern Kentucky, starting in Cincinnati. Formerly the CNO&TP subsidiary of Southern Railway, it is NS's most profitable line.
[84] http://www.census.gov/popest/counties/tables/CO-EST2005-02-21.csv
[85] Kentucky Counties (http://www.uky.edu/KentuckyAtlas/kentucky-counties.html), University of Kentucky
[88] Berea College: Learning, Labor, and Service (http://www.diversityweb.org/digest/vol10no1/mendel.cfm)
[92] [Abstract of A Guide to the Kentucky Education Reform Act of 1990 – provided by Education Resources Information Center (ERIC)]
[123] Jenifer K. Nii (2004). "Colonel's landmark KFC is mashed". Deseret Morning News. http://www.deseretnews.com/article/595057690/Colonels-landmark-KFC-is-mashed.html.Retrieved on October 28, 2007.
[129] http://kdla.ky.gov/resources/KYSymbols.htm

Bibliography

Politics

- Miller, Penny M. *Kentucky Politics & Government: Do We Stand United?* (1994) (http://www.questia.com/read/3096856/kentucky-politics-government-do-we-stand-united)
- Jewell, Malcolm E. and Everett W. Cunningham, *Kentucky Politics* (1968)

History

Surveys and reference

- Bodley, Temple and Samuel M. Wilson. *History of Kentucky* 4 vols. (1928).
- Caudill, Harry M., *Night Comes to the Cumberlands* (1963). ISBN 0-316-13212-8
- Channing, Steven. *Kentucky: A Bicentennial History* (1977).
- Clark, Thomas Dionysius. *A History of Kentucky* (many editions, 1937–1992).
- Collins, Lewis. *History of Kentucky* (1880).
- Harrison, Lowell H. and James C. Klotter. *A New History of Kentucky* (1997).
- Kleber, John E. et al. *The Kentucky Encyclopedia* (1992), standard reference history.
- Klotter, James C. *Our Kentucky: A Study of the Bluegrass State* (2000), high school text
- Lucas, Marion Brunson and Wright, George C. *A History of Blacks in Kentucky* 2 vols. (1992).
- Notable Kentucky African Americans http://web.archive.org/web/20080306060234/http://www.uky.edu/Subject/aakyall.html
- Share, Allen J. *Cities in the Commonwealth: Two Centuries of Urban Life in Kentucky* (1982).
- Wallis, Frederick A. and Hambleton Tapp. *A Sesqui-Centennial History of Kentucky* 4 vols. (1945).

- Ward, William S., *A Literary History of Kentucky* (1988) (ISBN 0-87049-578-X).
- WPA, *Kentucky: A Guide to the Bluegrass State* (1939) (http://www.questia.com/read/99066764/kentucky-a-guide-to-the-bluegrass-state), classic guide.

Specialized scholarly studies

- Bakeless, John. *Daniel Boone, Master of the Wilderness* (1989) (http://www.questia.com/read/105132587/daniel-boone-master-of-the-wilderness)
- Blakey, George T. *Hard Times and New Deal in Kentucky, 1929–1939* (1986)
- Coulter, E. Merton. *The Civil War and Readjustment in Kentucky* (1926)
- Davis, Alice. "Heroes: Kentucky's Artists from Statehood to the New Millennium" (2004)
- Ellis, William E. *The Kentucky River* (2000).
- Faragher, John Mack. *Daniel Boone* (1993)
- Fenton, John H. *Politics in the Border States: A Study of the Patterns of Political Organization, and Political Change, Common to the Border States: Maryland, West Virginia, Kentucky, and Missouri* (1957) (http://www.questia.com/read/6471305/politics-in-the-border-states-a-study-of-the-patterns)
- Ireland, Robert M. *The County in Kentucky History* (1976)
- Klotter, James C. *Kentucky: Portrait in Paradox, 1900–1950* (1992)
- Pearce, John Ed. *Divide and Dissent: Kentucky Politics, 1930–1963* (1987)
- Remini, Robert V. *Henry Clay: Statesman for the Union* (1991).
- Sonne, Niels Henry. *Liberal Kentucky, 1780–1828* (1939) (http://www.questia.com/read/100662336/liberal-kentucky-1780-1828)
- Tapp, Hambleton and James C Klotter. *Kentucky Decades of Discord, 1865–1900* (1977)
- Townsend, William H. *Lincoln and the Bluegrass: Slavery and Civil War in Kentucky* (1955) (http://www.questia.com/read/61648056/lincoln-and-the-bluegrass-slavery-and-civil-war-in)
- Waldrep, Christopher *Night Riders: Defending Community in the Black Patch, 1890–1915* (1993) (http://www.questia.com/read/14459929/night-riders-defending-community-in-the-black-patch) tobacco wars

External links

- Kentucky.gov: My New Kentucky Home (http://kentucky.gov/)
- Kentucky State Guide, from the Library of Congress (http://www.loc.gov/rr/program/bib/states/kentucky/index.html)
- Kentucky State and County Government Websites (http://www.genealogyinc.com/kentucky/government/)
- Kentucky (http://www.dmoz.org/Regional/North_America/United_States/Kentucky/) at the Open Directory Project
- Kentucky Department of Tourism (http://www.kentuckytourism.com/)
- The Kentucky Highlands Project (http://kentuckyhighlands.com/)
- USGS real-time, geographic, and other scientific resources of Kentucky (http://www.usgs.gov/state/state.asp?State=KY)
- Energy & Environmental Data for Kentucky (http://www.eia.gov/beta/state/?sid=KY)
- Kentucky State Facts (http://www.ers.usda.gov/data-products/state-fact-sheets/state-data.aspx?StateFIPS=21&StateName=Kentucky)
- Kentucky: Unbridled Spirit (http://kentucky.gov/about/Pages/unbridledspirit.aspx)
- Kentucky Virtual Library (http://www.kyvl.org/)
- U.S. Census Bureau Kentucky QuickFacts (http://quickfacts.census.gov/qfd/states/21000.html)
- Kentucky (http://ballotpedia.org/wiki/index.php/Kentucky) at Ballotpedia
- Kentucky State Databases (http://wikis.ala.org/godort/index.php/Kentucky) – Annotated list of searchable databases produced by Kentucky state agencies and compiled by the Government Documents Roundtable of the

American Library Association.

2000_United_States_Census

The **Twenty-second United States Census**, known as **Census 2000** and conducted by the Census Bureau, determined the resident population of the United States on April 1, 2000, to be 281,421,906, an increase of 13.2% over the 248,709,873 persons enumerated during the 1990 Census.[1] This was the twenty-second federal census and was at the time the largest civility administrative peacetime effort in the United States.[2]

Approximately 16 percent of households received a "long form" of the 2000 census, which contained over 100 questions. Full documentation on the 2000 census, including census forms and a procedural history, is available from the Integrated Public Use Microdata Series.

Data availability

Microdata from the 2000 census is freely available through the Integrated Public Use Microdata Series. Aggregate data for small areas, together with electronic boundary files, can be downloaded from the National Historical Geographic Information System.

State rankings

Rank	State	Population as of 1990 Census[]	Population as of 2000 Census[]	Change	Percent change
1	California	29,760,021	33,871,648	▲ 4,111,627	▲ 13.8%
2	Texas	16,986,510	20,851,820	▲ 3,865,510	▲ 22.8%
3	New York	17,990,455	18,976,457	▲ 986,002	▲ 5.5%
4	Florida	12,937,926	15,982,378	▲ 3,044,452	▲ 23.5%
5	Illinois	11,430,602	12,419,293	▲ 988,691	▲ 8.6%
6	Pennsylvania	11,881,643	12,281,054	▲ 399,411	▲ 3.4%
7	Ohio	10,847,115	11,353,140	▲ 506,025	▲ 4.7%
8	Michigan	9,295,297	9,938,444	▲ 643,147	▲ 6.9%
9	New Jersey	7,730,188	8,414,350	▲ 684,162	▲ 8.9%
10	Georgia	6,478,216	8,186,453	▲ 1,708,237	▲ 26.4%
11	North Carolina	6,628,637	8,049,313	▲ 1,420,676	▲ 21.4%
12	Virginia	6,187,358	7,078,515	▲ 891,157	▲ 14.4%
13	Massachusetts	6,016,425	6,349,097	▲ 332,672	▲ 5.5%
14	Indiana	5,544,159	6,080,485	▲ 536,326	▲ 9.7%
15	Washington	4,866,692	5,894,121	▲ 1,027,429	▲ 21.1%
16	Tennessee	4,877,185	5,689,283	▲ 812,098	▲ 16.7%
17	Missouri	5,117,073	5,595,211	▲ 478,138	▲ 9.3%
18	Wisconsin	4,891,769	5,363,675	▲ 471,906	▲ 9.6%
19	Maryland	4,781,468	5,296,486	▲ 515,018	▲ 10.8%

20	**Arizona**	3,665,228	5,130,632	▲ 1,465,404	▲ 40.0%
21	**Minnesota**	4,375,099	4,919,479	▲ 544,380	▲ 12.4%
22	**Louisiana**	4,219,973	4,468,976	▲ 249,003	▲ 5.9%
23	**Alabama**	4,040,587	4,447,100	▲ 406,513	▲ 10.1%
24	**Colorado**	3,294,394	4,301,261	▲ 1,006,867	▲ 30.6%
25	**Kentucky**	3,685,296	4,041,769	▲ 356,473	▲ 9.7%
26	**South Carolina**	3,486,703	4,012,012	▲ 525,309	▲ 15.1%
27	**Oklahoma**	3,145,585	3,450,654	▲ 305,069	▲ 9.7%
28	**Oregon**	2,842,321	3,421,399	▲ 579,078	▲ 20.4%
29	**Connecticut**	3,287,116	3,405,565	▲ 118,449	▲ 3.6%
30	**Iowa**	2,776,755	2,926,324	▲ 149,569	▲ 5.4%
31	**Mississippi**	2,573,216	2,844,658	▲ 271,442	▲ 10.5%
32	**Kansas**	2,477,574	2,688,418	▲ 210,844	▲ 8.5%
33	**Arkansas**	2,350,725	2,673,400	▲ 322,675	▲ 13.7%
34	**Utah**	1,722,850	2,233,169	▲ 510,319	▲ 29.6%
35	**Nevada**	1,201,833	1,998,257	▲ 796,424	▲ 66.3%
36	**New Mexico**	1,515,069	1,819,046	▲ 303,977	▲ 20.1%
37	**West Virginia**	1,793,477	1,808,344	▲ 14,867	▲ 0.8%
38	**Nebraska**	1,578,385	1,711,263	▲ 132,878	▲ 8.4%
39	**Idaho**	1,006,749	1,293,953	▲ 287,204	▲ 28.5%
40	**Maine**	1,227,928	1,274,923	▲ 46,995	▲ 3.8%
41	**New Hampshire**	1,109,252	1,235,786	▲ 126,534	▲ 11.4%
42	**Hawaii**	1,108,229	1,211,537	▲ 103,308	▲ 9.3%
43	**Rhode Island**	1,003,464	1,048,319	▲ 44,855	▲ 4.5%
44	**Montana**	799,065	902,195	▲ 103,130	▲ 12.9%
45	**Delaware**	666,168	783,600	▲ 117,432	▲ 17.6%
46	**South Dakota**	696,004	754,844	▲ 58,840	▲ 8.5%
47	**North Dakota**	638,800	642,200	▲ 3,400	▲ 0.5%
48	**Alaska**	550,043	626,932	▲ 76,889	▲ 14.0%
49	**Vermont**	562,758	608,827	▲ 46,069	▲ 8.2%
—	**District of Columbia**	606,900	572,059	▼ -34,841	▼ -5.7%
50	**Wyoming**	453,588	493,782	▲ 40,194	▲ 8.9%
	United States	**248,709,873**	**281,421,906**	**32,712,033**	**13.2%**

Population profile

The U.S. resident population includes the total number of people in the 50 states and the District of Columbia. The Bureau also enumerated the residents of the associated state of Puerto Rico; its population was 3,808,610, an 8.1% increase over the number from a decade earlier.

In an introduction to a more detailed population profile (see references below), the Census Bureau highlighted the following facts about U.S population dynamics:

- 75% of respondents said they were White or Caucasian and no other race;
- Hispanics accounted for 12.5% of the U.S. population, up from 9% in 1990;
- 12.3% (34.5 million Americans) were of German descent; German Americans
- 12.3% were of Black or African American descent;
- 3.6% of respondents were Asian;
- 2.4% (6.8 million Americans)[3] of respondents were multiracial (2 or more races). The 2000 Census was the first time survey options for multiracial Americans were provided.
- Between 1990 and 2000, the population aged 45 to 54 grew by 49% and those aged 85 and older grew 38%;
- Women outnumbered men two to one among those aged 85 and older;
- Almost one in five adults had some type of disability in 1997 and the likelihood of having a disability increased with age;
- Families (as opposed to men or women living alone) still dominated American households, but less so than they did thirty years ago;
- Since 1993, both families and non-families have seen median household incomes rise, with "households headed by a woman without a spouse present" growing the fastest;
- People in married-couple families had the lowest poverty rates;
- The poor of any age were more likely than others to lack health insurance coverage;
- The number of elementary and high school students in 2000 fell just short of the all-time high of 49 million reached in 1970;
- Improvements in educational attainment cross racial and ethnic lines; and
- The majority (51%) of U.S. households had access to computers; 42% have Internet access.[4]

Changes in population

Regionally, the South and West experienced the bulk of the nation's population increase, 14,790,890 and 10,411,850, respectively. This meant that the mean center of U.S. population moved to Phelps County, Missouri. The Northeast grew by 2,785,149; the Midwest by 4,724,144.

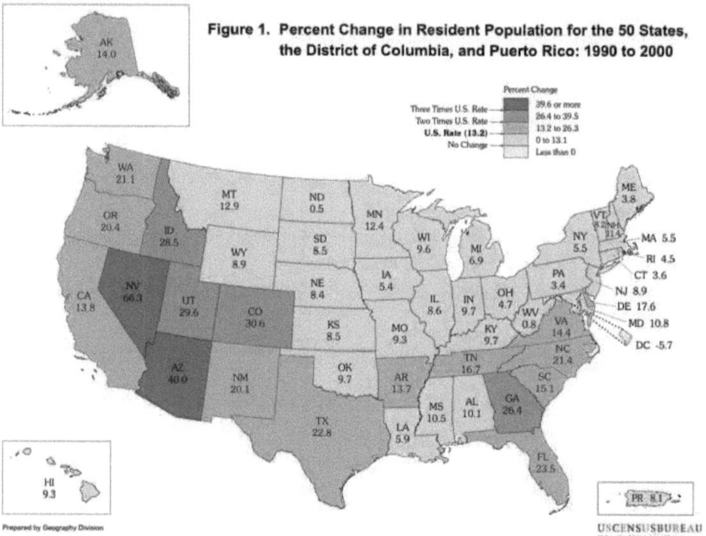

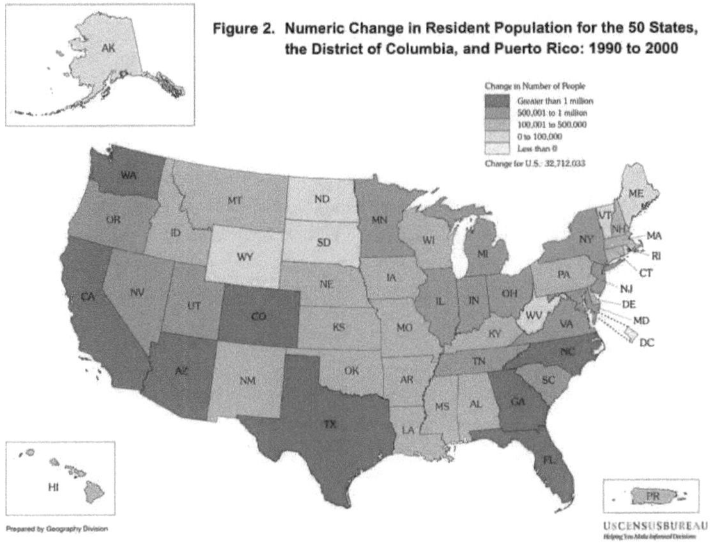

(maps not to scale)

Reapportionment

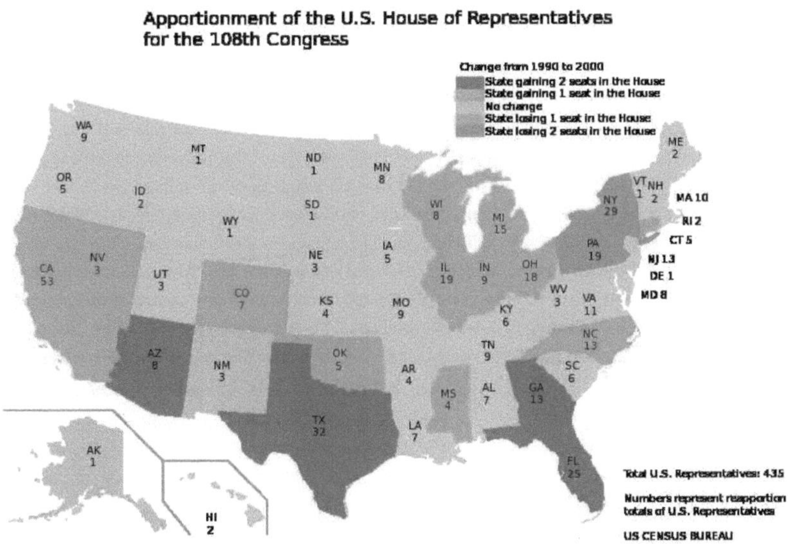

The results of the census are used to determine how many congressional districts each state is apportioned. Congress defines the formula, in accordance with Title 2 of the U.S. Code, to reapportion among the states the 435 seats in the United States House of Representatives. The apportionment population consists of the resident population of the fifty states, plus the overseas military and federal civilian employees and their dependents living with them who could be allocated to a state. Each member of the House represents a population of about 647,000. The populations of the District of Columbia and Puerto Rico are excluded from the apportionment population because they do not have voting seats in the U. S. House of Representatives.

Since the first census in 1790, the decennial count has been the basis for the United States representative form of government. Article I, Section II specifies that "The Number of Representatives shall not exceed one for every thirty Thousand, but each State shall have at Least one Representative." In 1790, each member of the House represented about 34,000 residents. Since then, the House more than quadrupled in size, and in 1911 the number of representatives was fixed at 435. Today, each member represents about 20 times as many constituents.

Adjustment controversy

In the years leading up to the 2000 census, there was substantial controversy over whether the Bureau should adjust census figures based on a follow-up survey, called the post-enumeration survey, of a sample of blocks. (In 1999, the Supreme Court ruled 5-4 that the Constitution prohibits the use of such figures for apportionment purposes, but it may be permissible for other purposes where feasible.) The controversy was partly technical, but also partly political, since based on data from the 1990 census both parties believed that adjustment would likely have the effect, after redistricting, of slightly increasing Democratic representation in legislative bodies, but would also give Utah an additional, probably Republican, representative to Congress.[5] [6]

Following the census, discrepancies between the adjusted census figures and demographic estimates of population change could not be resolved in time to meet legal deadlines for the provision of redistricting data, and the Census Bureau therefore recommended that the unadjusted results be used for this purpose.[7] This recommendation was followed by the Secretary of Commerce (the official in charge of making the determination).

Utah controversy

After the census was tabulated, Utah challenged the results in two different ways. Utah was extremely close to gaining a fourth congressional seat, falling 857 people short, which in turn was allocated to North Carolina. The margin was later shortened to 80 people, after the federal government discovered that it overcounted the population of North Carolina by 2,673 residents.[8] The Census Bureau counted members of the military and other federal civilian employees serving abroad as residents of their home state but did not count other individuals living outside the United States. Utah claimed that individuals traveling abroad as religious missionaries should be counted as residents and that the failure to do so imposed a burden on Mormon religious practice. Almost half of all Mormon missionaries, more than 11,000 individuals, were from Utah; only 102 came from North Carolina. If this policy were changed, then Utah would have received an additional seat instead of North Carolina. On November 26, 2002, the Supreme Court affirmed the lower court ruling that rejected Utah's efforts to have Mormon missionaries counted.[9]

The state of Utah then filed another lawsuit alleging that the statistical methods used in computing the state populations were improper and cost Utah the seat. The Bureau uses a method called imputation to assign a number of residents to addresses where residents cannot be reached after multiple efforts. While nationwide the imputation method added .4% to the population, the rate in Utah was .2%. The state challenged that the use of imputation violates the Census Act of 1957 and that it also fails the Constitution's requirement in Article I, Section 2 that an "actual enumeration" be used for apportionment.[10] This case, *Utah v. Evans*, made it to the Supreme Court, but Utah was again defeated.[11]

Gay and lesbian controversy

The census forms did not include any questions regarding sexual orientation, making it impossible to compile data comparing heterosexual and homosexual populations. However, two questions *were* asked that allowed same-sex partnerships to be counted. The questionnaires asked the sex of each person in a household and they asked what the relationship was between each of the members of the household. Respondents could check "Husband/wife" or "unmarried partner" or a number of other relationships.[12] [13] Responses were tabulated and the Census Bureau reported that there were more than 658,000 same-sex couples heading households in the United States. However, only about 25% of gay men and 40% of lesbians are in shared-household partnerships at any one time, according to non-Census surveys.[14] For every same-sex couple tallied in the census, there could be three to six more homosexual un-partnered individuals who would not be counted as gay. The Census reported that same-sex male couples numbered 336,001 and female same-sex couples numbered 329,522.[15] Extrapolating from those figures and the surveyed partnering habits of homosexuals, as many as 4.3 million homosexual adults could have been living in the U.S. in 2000. The exact number cannot be known because the Census did not count them specifically. Bisexual and transgender populations were not counted, either, because there were no questions regarding this information. Also unavailable is the number of additional same-sex couples living under the same roof as the first, though this applies to additional heterosexual couples as well. The lack of accurate numbers makes it difficult for lawmakers who are considering legislation on hate crimes or social services for gay families with children.[16] It also makes for less accuracy when predicting the fertility of a population.[]

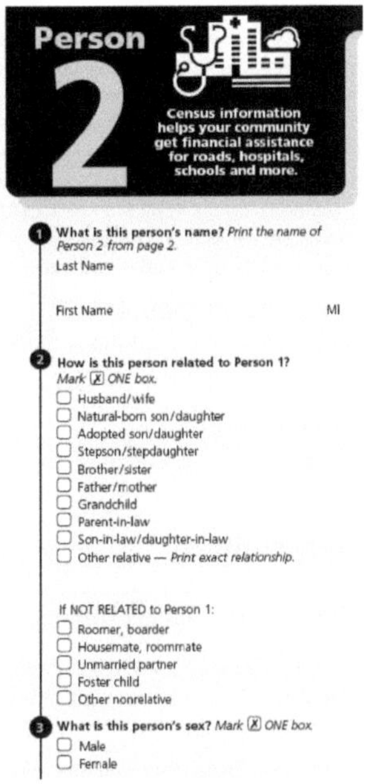
Person 2

Census information helps your community get financial assistance for roads, hospitals, schools and more.

1 What is this person's name? *Print the name of Person 2 from page 2.*

Last Name

First Name MI

2 How is this person related to Person 1? *Mark ☒ ONE box.*

- ☐ Husband/wife
- ☐ Natural-born son/daughter
- ☐ Adopted son/daughter
- ☐ Stepson/stepdaughter
- ☐ Brother/sister
- ☐ Father/mother
- ☐ Grandchild
- ☐ Parent-in-law
- ☐ Son-in-law/daughter-in-law
- ☐ Other relative — *Print exact relationship.*

If NOT RELATED to Person 1:

- ☐ Roomer, boarder
- ☐ Housemate, roommate
- ☐ Unmarried partner
- ☐ Foster child
- ☐ Other nonrelative

3 What is this person's sex? *Mark ☒ ONE box.*

- ☐ Male
- ☐ Female

Census 2000 Long Form Questionnaire showing the Person **2** section including questions *2* and *3* which allow data to be compiled regarding same-sex partners

Another issue that concerned gay rights advocates involved the automatic changing of data during the tabulation process. This automatic software data compiling method, called *allocation*, was designed to counteract mistakes and discrepancies in returned questionnaires. Forms that were filled out by two same-sex persons who checked the "Husband/wife" relationship box were treated as a discrepancy. The Census Bureau explained that same-sex "Husband/wife" data samples were changed to "unmarried partner" by computer processing methods in 99% of the cases. In the remaining 1%, computer systems used one of two possibilities: a) one of the two listed sexes was changed, making the partnership appear heterosexual, or b) if the two partners were more than 15 years apart in age, they might have been reassigned into a familial parent/child relationship.[] The process of automatic reassignment of same-sex marriage data was initiated so that the Census Bureau would not contravene the Defense of Marriage Act passed in 1996. The Act states:

> In determining the meaning of any Act of Congress, or of any ruling, regulation or interpretation of the various administrative bureaus and agencies of the United States, the word 'marriage' means only a legal union between one man and one woman as husband and wife, and the word 'spouse' refers only to a person of the opposite sex who is a husband or wife.[]

With allocation moving married same-sex couples to the unmarried partner category, social scientists lost information that could have been extracted relating to the social stability of a same gender couple who identify themselves as married.[]

References

[3] Jayson, Sharon. "'Colorblind' Generation Doesn't Blink at interracial Relationships." USA TODAY. Feb. 7 2006: n.p. SIRS Researcher. Web. 25 Oct 2010.

[6] (http://www.brook.edu/comm/policybriefs/pb56.htm)

[9] Greenhouse, Linda. "Justices Deal Utah a Setback In Its Bid to Gain a House Seat" (http://query.nytimes.com/gst/fullpage.html?res=940DE5DF1E3AF934A15752C1A9679C8B63&scp=1&sq=utah mormon abroad census&st=cse), *The New York Times*, November 27, 2001. Accessed July 16, 2008.

[10] Greenhouse, Linda. "Supreme Court Roundup; Justices to Hear Utah's Challenge to Procedure in 2000 Census" (http://query.nytimes.com/gst/fullpage.html?res=9904E3DB103BF930A15752C0A9649C8B63), *The New York Times*, January 23, 2002. Accessed July 16, 2008.

[11] Greenhouse, Linda. "THE SUPREME COURT: RIGHT TO PRIVACY; Supreme Court Finds Law On Educational Privacy Isn't Meant for Individuals" (http://query.nytimes.com/gst/fullpage.html?res=9B03E6D7123FF932A15755C0A9649C8B63&sec=&spon=&pagewanted=2), *The New York Times*, June 21, 2002. Accessed July 16, 2008.

- Constitution Article I Section II re Enumeration and Apportionment (http://www.senate.gov/civics/constitution_item/constitution.htm#a1_sec1)

Further reading

- .

External links

United States Census Bureau

- Census 2000 gateway (http://www.census.gov/main/www/cen2000.html)
- Population Profile of the United States: 2000 (http://www.census.gov/population/www/pop-profile/profile2000.html)
- Population Profile Introductory slide show (http://www.census.gov/population/pop-profile/2000/profile2000.ppt), in MS Powerpoint format
- State and County QuickFacts (http://quickfacts.census.gov/qfd/), the most requested information
- American FactFinder (http://factfinder.census.gov), for population, housing, economic, and geographic data

Other 2000 census websites

- MLA Language Map (http://www.mla.org/census_main) from the Modern Language Association
- How the Census Works (http://people.howstuffworks.com/census.htm) via Howstuffworks.com
- step-by-step how to analyze the 2000 US Census Public Use Microdata Sample (PUMS) with free tools website (http://www.asdfree.com/search/label/united states decennial census public use microdata sample (pums))

United_States_Census_Bureau

Bureau of the Census	
Seal	
Logo	
Agency overview	
Formed	July 1, 1903
Preceding Agency	Temporary census offices
Headquarters	Suitland, Maryland
Employees	5,593 (2006)
Annual budget	US$3.1 billion (2009) US$7.2 billion (est. 2010) US$1.3 billion (est. 2011)
Agency executive	Director, Thomas L. Mesenbourg (acting)
Parent agency	Economics and Statistics Administration
Website	
www.census.gov [1]	

The **United States Census Bureau** (officially the **Bureau of the Census**, as defined in Title 13 U.S.C. § 11 [2]) is a principal agency of the U.S. Federal Statistical System responsible for producing data about the American people and economy.

The primary mission of the Census Bureau is conducting the U.S. Census every ten years, which allocates the seats of the U.S. House of Representatives to the states based on their population.[3] In addition to the decennial census, the Census Bureau continually conducts dozens of other censuses and surveys, including the American Community Survey, the U.S. Economic Census, and the Current Population Survey.[3] Furthermore, economic and foreign trade indicators released by the federal government typically contain data produced by the Census Bureau. The various censuses and surveys conducted by the Census Bureau help allocate over $400 billion in federal funds every year and help states, local communities, and businesses make informed decisions.[]

The Census Bureau is situated within the U.S. Department of Commerce and its director is appointed by the President of the United States.

Legal mandate

The Constitution of the United States (Article I, section II) directs that the population be enumerated at least once every ten years and the resulting counts used to set the number of members from each state in the House of Representatives and, by extension, in the Electoral College. The Census Bureau now conducts a full population count every 10 years in years ending with a 0 (zero) and uses the term "decennial" to describe the operation. Between censuses, the Census Bureau makes population estimates and projections.[4]

In addition, Census data directly affects how more than $400 billion per year in federal and state funding is allocated to communities for neighborhood improvements, public health, education, transportation and much more. The Census Bureau is mandated with fulfilling these obligations: the collecting of statistics about the nation, its people, and economy. The Census Bureau's legal authority is codified in Title 13 of the United States Code.

The Census Bureau also conducts surveys on behalf of various federal government and local government agencies on topics such as employment, crime, health, consumer expenditures, and housing. Within the bureau, these are known as "demographic surveys" and are conducted perpetually between and during decennial (10-year) population counts. The Census Bureau also conducts economic surveys of manufacturing, retail, service, and other establishments and of domestic governments.

Between 1790 and 1840, the census was taken by marshals of the judicial districts.[5] The Census Act of 1840 established a central office[6] which became known as the Census Office. Several acts followed that revised and authorized new censuses, typically at the 10-year intervals. In 1902 the temporary Census Office was moved under the Department of Interior, and in 1903 it was renamed the Census Bureau under the new Department of Commerce and Labor. The department was intended to consolidate overlapping statistical agencies, but Census Bureau officials were hindered by their subordinate role in the department.[7]

An act in 1920 changed the date and authorized manufacturing censuses every 2 years and agriculture censuses every 10 years.[8] In 1929, a bill was passed mandating that the House of Representatives be reapportioned based on the results of the 1930 Census.[8] In 1954, various acts were codified into Title 13 of the US Code.[9]

By law the Census Bureau must count everyone and submit state population totals to the U.S. President by December 31 of any year ending in a zero. States within the Union receive the results in the spring of the following year.

Uses of census data

Many federal, state, local and tribal governments use census data to:

- Decide the location of new housing and public facilities,
- Examine the demographic characteristics of communities, states, and the USA,
- Plan transportation systems and roadways,
- Determine quotas and creation of police and fire precincts, and
- Create localized areas for elections, schools, utilities, etc.

Data stewardship

The United States Census Bureau is committed to confidentiality, and guarantees non-disclosure of any addresses or personal information related to individuals or establishments. Title 13 of the U.S. Code establishes penalties for the disclosure of this information. All Census employees must sign a sworn affidavit of non-disclosure prior to employment.

The Bureau cannot share responses, addresses or personal information with anyone including United States or foreign government and law enforcement agencies such as the IRS or the FBI or Interpol. "Providing quality data, for public good—while respecting individual privacy and, at the same time, protecting confidentiality—is the Census Bureau's core responsibility", "Keeping the public's trust is critical to the Census's ability to carry out the mission as the leading source of quality data about the Nation's people and economy."[10] Only after 72 years does the information collected become available to other agencies or the general public.

In 1918, the Census Bureau released individual information regarding several hundred young men to the Justice Department and Selective Service system for the purpose of prosecutions for draft evasion.[11] [12] During World War II, the United States Census Bureau assisted the government's Japanese American internment efforts by providing confidential neighborhood information on Japanese-Americans. The Bureau's role was denied for decades but was finally proven in 2007.[13] [14]

Organizational structure

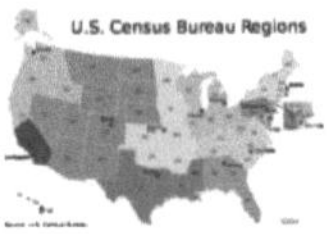

Pre-2013

Post-2013

Since 1903, the official census-taking agency of the United States government has been the Bureau of the Census. The Census Bureau is headed by a Director, assisted by a Deputy Director and an Executive Staff composed of the associate directors.

Census headquarters in Suitland, Maryland

The Census Bureau has had headquarters in Suitland, Maryland, since 1942. A new headquarters complex was completed in 2007 and supports over 4,000 employees.[15] The Bureau operates regional offices in 12 cities: Boston, New York City, Philadelphia, Detroit, Chicago, Kansas City, Seattle, Charlotte, Atlanta, Dallas, Denver, and Los Angeles. The National Processing Center is located in Jeffersonville, Indiana. Additional temporary processing facilities are used to facilitate the decennial census, which employs more than a million people. The cost of the 2000 Census was $4.5 billion. During the years just prior to the decennial census, parallel census offices, known as "Regional Census Centers" are opened in the field office cities. The decennial operations are carried out from these facilities. The Regional Census Centers oversee the openings and closings of smaller "Local Census Offices" within their collection jurisdictions. The estimated cost of the 2010 Census is $14.7 billion.

On January 1, 2013, the Census Bureau was to consolidate its 12 regional offices into 6. Increasing costs of data collection, changes in survey management tools such as laptops and the increasing use of multi-modal surveys (i.e. internet, telephone, and in-person) has led the Census Bureau to consolidate.[16] The remaining regional offices are to be located in: New York City, Philadelphia, Chicago, Atlanta, Denver, and Los Angeles.[17]

The Census Bureau also runs the Census Information Center cooperative program that involves 58 "national, regional, and local non-profit organizations." The CIC program aims to represent the interests of underserved communities.[18]

Census regions and divisions

The United States Census Bureau has four official regions, with nine official divisions.[19]

- Northeast
  - New England
  - Middle Atlantic
- Midwest
  - East North Central States
  - West North Central States
- West
  - Pacific States
  - Mountain States
- South
  - West South Central States
  - East South Central States
  - South Atlantic States

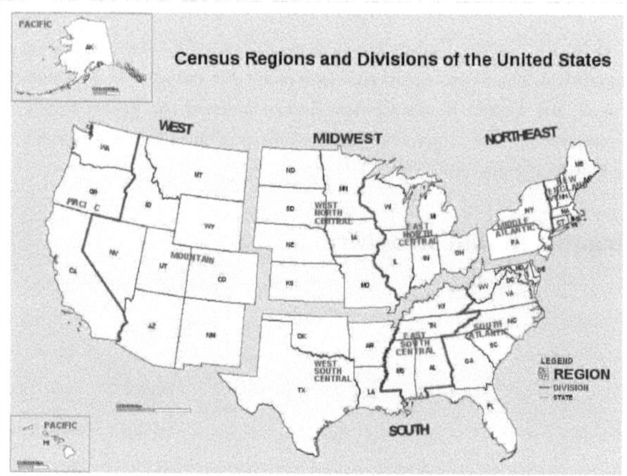

Map of official Census regions and divisions.

Computer equipment

A card puncher, part of the tabulation system used to compile the thousands of facts gathered by the Bureau. Holes are punched in the card according to a prearranged code transferring the facts from the census questionnaire into statistics.

The 1890 census was the first to use the electric tabulating machines invented by Herman Hollerith.[20] [21] For 1890–1940 details, see In 1946, knowing of the Bureau's funding of Hollerith and, later, Powers, John Mauchly approached the Bureau about early funding for UNIVAC development.[22] A UNIVAC I computer was accepted by the Bureau in 1951.[23]

In order to reduce paper usage and reduce payroll expenses, 500,000 handheld computers (HHC's) were used for the first time in 2009 during the address canvassing portion of the 2010 Decennial Census Project. Projected savings were estimated to be over $1 billion.[24]

Ongoing surveys

A survey is a method of collecting and analyzing social, economic, and geographic data. It provides information about the conditions of the United States, states, and counties. Throughout the decade between censuses, the bureau is conducts surveys to produce a general view and comprehensive study of the United States' social and economic conditions.

Staff from the Current Surveys Program conduct ongoing and special surveys about people and their characteristics. A network of professional field representatives gathers information from a sample of households, responding to questions about employment, consumer expenditures, health, housing, and other topics. Surveys conducted between decades:

- American Community Survey
- American Housing Survey
- Consumer Expenditure Survey
- Census of Governments
- Current Population Survey
- Economic Census
- National Hospital Discharge Survey
- National Health Interview Survey
- National Hospital Ambulatory Medical Care Survey
- National Crime Victimization Survey
- National Ambulatory Medical Care Survey (NAMCS)
- National Nursing Home Survey
- Survey of Income and Program Participation
- Survey of Construction
- Survey of Market Absorption
- Survey of Program Dynamics
- National Longitudinal Survey
- National Survey of Fishing, Hunting, & Wildlife-Associated Recreation 2001
- American Housing Survey
- Residential Finance Survey
- National Epidemiologic Survey of Alcohol Related Conditions
- Annual Retail Trade Survey
- Annual Wholesale Trade Survey
- Annual and Quarterly Services Surveys

Other surveys conducted

The Census Bureau collects information in many other surveys and provides the data to the survey sponsor for release. These sponsors include:

- Bureau of Justice Statistics (BJS)
- Bureau of Labor Statistics (BLS)
- Bureau of Transportation Statistics (BTS)
- Department of Housing and Urban Development (HUD)
- National Center for Education Statistics (NCES)
- National Center for Health Statistics (NCHS)
- National Science Foundation (NSF)
- Social Security Administration (SSA)

See also

- Outline of the United States
- Index of United States-related articles
- Book:United States
- Geography
  - Human geography
    - Demography
- United States
- Demographics of the United States
  - United States Census Bureau
    - List of U.S. states and territories by population
    - List of metropolitan areas of the United States
    - List of United States cities by population
    - List of United States counties and county-equivalents
  - United States Office of Management and Budget
    - The OMB has defined 1098 statistical areas comprising 388 MSAs, 541 μSAs, and 169 CSAs
      - Primary statistical area – List of the 574 PSAs
        - Combined Statistical Area – List of the 169 CSAs
        - Core Based Statistical Area – List of the 929 CBSAs
          - Metropolitan Statistical Area – List of the 388 MSAs
          - Micropolitan Statistical Area – List of the 541 μSAs

The 29 US MDs

- Micropolitan Statistical Area

The 576 US μSAs

- United States urban area

List of United States urban areas

Notes

[1] http://www.census.gov
[2] http://www.law.cornell.edu/uscode/13/1.html
[3] USCB DOC-D1026 QVC Manual 01/03/09
[5] History 1790 (http://www.census.gov/history/www/overview/012270.html). US Census Bureau.
[6] History 1840 (http://www.census.gov/history/www/overview/012275.html). US Census Bureau.
[7] History: 1900 Overview (http://www.census.gov/history/www/through_the_decades/overview/1900.html). US Census Bureau.
[8] History 1920 (http://www.census.gov/history/www/overview/012283.html). US Census Bureau.
[9] History 1954 (http://www.census.gov/history/www/overview/012286.html). US Census Bureau.
[15] Census.gov (http://www.census.gov/aboutus/who.html)
[20] Herman Hollerith (http://www.census.gov/history/www/notable_alumni/010944.html)
[21] History 1890 (http://www.census.gov/history/www/overview/012280.html) US Census Bureau.
[24] Govcomm.harris.com (http://www.govcomm.harris.com/solutions/products/000138.asp)

References

- The original version of, and subsequent additions to, this article were adapted from U.S. Census Bureau (http://www.census.gov/acsd/www/history.html) text.
- Geographic Areas Reference Manual (http://www.census.gov/geo/www/garm.html) from the U.S. Census Bureau contains detailed explanations of geographic terms used in the census.
- The mission and operation of the U.S. Census Bureau are laid out in Title 13 of the United States Code.

External links

- United States Census Bureau (http://www.census.gov/)
  - USCB population estimates (http://www.census.gov/popest/data/index.html)
  - USCB History (http://www.census.gov/history)
  - U.S. and World Population Clocks – POPClocks (http://www.census.gov/main/www/popclock.html)
  - Works by Bureau of the Census (http://www.gutenberg.org/author/United_States._Bureau_of_the_Census) at Project Gutenberg
  - USCB Meeting Notices and Rule Changes (http://thefederalregister.com/b.p/agency/Census_Bureau/) from The Federal Register RSS Feed (http://thefederalregister.com/rss/agency/Census_Bureau/)
  - Proposed and finalized federal regulations from the USCB (http://openregs.com/agencies/view/173/bureau_of_the_census)

Census

A **census** is the procedure of systematically acquiring and recording information about the members of a given population. It is a regularly occurring and official count of a particular population.[1] The term is used mostly in connection with national population and housing censuses; other common censuses include agriculture, business, and traffic censuses. The United Nations defines the essential features of population and housing censuses as "individual enumeration, universality within a defined territory, simultaneity and defined periodicity", and recommends that population censuses be taken at least every 10 years. United Nations recommendations also cover census topics to be collected, official definitions, classifications and other useful information to coordinate international practice.[2] [3]

Census taker visits a family living in a caravan, Netherlands 1925

The word is of Latin origin; during the Roman Republic, the census was a list that kept track of all adult males fit for military service. The modern census is essential to international comparisons of any kind of statistics and censuses collect data on many attributes of the population, not just how many people there are, although population estimates remain an important function of the census.

The census can be contrasted with sampling in which information is obtained only from a subset of a population, sometimes as an intercensal estimate. Census data are commonly used for research, business marketing, and planning, as well as a baseline for sampling surveys. Census counts are necessary to adjust samples to be representative of a population by weighting them as is common in opinion polling. Similarly, stratification requires knowledge of the relative sizes of different population strata which are derived from census enumerations. In some countries, census data are used to apportion electoral representation (sometimes controversially – e.g., *Utah v. Evans*).

Sampling

A census is often construed as the opposite of a sample as everyone is surveyed rather than a fraction. However, population census relies on a sampling frame to count the population. This is the only way to be sure that everyone has been included as otherwise those not responding would not be followed up and individuals could be missed. The fundamental premise of a census is that the population is not known and a new estimate is to be made by the analysis of primary data.

The use of a sampling frame is counterintuitive as it suggests that the population size is already known. However, a census is also used to collect attribute data on the individuals in the nation. This process of sampling marks the difference between historical census, which was a house to house process or the product of an imperial decree, and the modern statistical project.

The sampling frame used by census is almost always an address register. Thus it is not known if there is anyone resident or how many people there are in each household. Depending on the mode of enumeration, a form is sent to the householder, an enumerator calls, or administrative records for the dwelling are accessed. As a preliminary to the dispatch of forms, census workers will check any address problems on the ground.

While it may seem straightforward to use the postal service file for this purpose, this can be out of date and some dwellings may contain a number of independent households. A particular problem is what are termed 'communal establishments' which may be student residences, religious orders, homes for the elderly etc. As these are not easily enumerated by a single householder, they are often treated differently and visited by special teams of census workers to make sure they are classified appropriately.

Residence definitions

Individuals are normally counted within households and information is typically collected about the household structure and the housing. For this reason international documents refer to censuses of population and housing. Normally the census response is made by a household, indicating details of individuals resident there.

An important aspect of census enumerations is determining which individuals can be counted from which cannot be counted. Broadly, three definitions can be used: *de facto* residence; *de jure* residence; and, permanent residence. This is important to consider individuals who have multiple or temporary addresses. Every person should be identified uniquely as resident in one place but where they happen to be on Census Day, their *de facto* residence, may not be the best place to count them. Where an individual uses services may be more useful and this is at their usual, or *de jure*, residence. An individual may be represented at a permanent address, perhaps a family home for students or long term migrants.

It is necessary to have a precise definition of residence to decide whether visitors to a country should be included in the population count. This is becoming more important as students travel abroad for education for a period of several years. Other groups causing problems of enumeration are new born babies, refugees, people away on holiday, people moving home around census day, and people without a fixed address.

People having second homes because of working in another part of the country or retaining a holiday cottage are difficult to fix at a particular address sometimes causing double counting or houses being mistakenly identified as vacant. Another problem is where people use a different address at different times e.g. students living at their place of education in term time but returning to a family home during vacations or children whose parents have separated who effectively have two family homes. Census enumeration has always been based on finding people where they live as there is no systematic alternative - any list you could use to find people is derived from census activities in the first place. Recent UN guidelines provide recommendation on enumerating such complex households.[4]

Enumeration strategies

Historical censuses used crude enumeration assuming absolute accuracy. Modern approaches take into account the problems of overcount and undercount, and the coherence of census enumerations with other official sources of data.[5] This reflects a realist approach to measurement, acknowledging that under any definition of residence there is a true value of the population but this can never be measured with complete accuracy. An important aspect of the census process is to evaluate the quality of the data.[6]

Many countries use a post-enumeration survey to adjust the raw census counts.[7] This works in a similar manner to capture-recapture estimation for animal populations. In census circles this method is called dual system enumeration (DSE). A sample of households are visited by interviewers who record the details of the household as at census day. These data are then matched to census records and the number of people missed can be estimated by considering the number missed in the census or survey but counted in the other. This way counts can be adjusted for non-response varying between different demographic groups. An explanation using a fishing analogy can be found in "Trout, Catfish and Roach..."[8] which won an award from the Royal Statistical Society for excellence in official statistics in 2011.

Triple system enumeration has been proposed as an improvement as it would allow evaluation of the statistical dependence of pairs of sources. However, as the matching process is the most difficult aspect of census estimation this has never been implemented for a national enumeration. It would also be difficult to identify three different sources that were sufficiently different to make the triple system effort worthwhile. The DSE approach has another weakness in that it assumes there is no person counted twice (over count). In *de facto* residence definitions this would not be a problem but in *de jure* definitions individuals risk being recorded on more than one form leading to double counting. A particular problem here are students who often have a term time and family address.

Several countries have used a system which is known as short form/long form.[9] This is a sampling strategy which randomly chooses a proportion of people to send a more detailed questionnaire to (the long form). Everyone receives the short form questions. Thereby more data are collected but not imposing a burden on the whole population. This also reduces the burden on the statistical office. Indeed in the UK all residents were required to fill in the whole form but only a 10% sample were coded and analysed in detail, until 2001.[10] New technology means that all data are now scanned and processed. Recently there has been controversy in Canada about the cessation of the long form with the head, Munir Sheikh resigning.[11]

The use of alternative enumeration strategies is increasing[12] but these are not so simple as many people assume. The Netherlands has been most advanced in adopting a census using administrative data. This allows a simulated census to be conducted by linking several different administrative databases at an agreed time. Data can be matched and an overall enumeration established accounting for where the different sources are discrepant. A validation survey is still conducted in a similar way to the post enumeration survey employed in a traditional census.

Other countries which have a population register use this as a basis for all the census statistics needed by users. This is most common amongst Nordic countries but requires a large number of different registers to be combined including population, housing, employment and education. These registers are then combined and brought up to the standard of a statistical register by comparing the data in different sources and ensuring the quality is sufficient for official statistics to be produced.[13]

A recent innovation is the French instigation of a rolling census programme with different regions enumerated each year such that the whole country is completely enumerated every 5 years.

In Europe, in connection with the 2010 census round, a large number of countries adopted alternative census methodologies, often based on the combination of data from registers, surveys and other sources.[14]

Technology

Censuses have evolved in their use of technology with the latest censuses, the 2010 round, using many new types of computing. In Brazil, handheld devices were used by enumerators to locate residences on the ground. In many countries, census returns could be made via the Internet as well as in paper form. DSE is facilitated by computer matching techniques which can be automated, such as propensity score matching. In the UK, all census formats are scanned and stored electronically before being destroyed, replacing the need for physical archives. The record linking to perform an administrative census would not be possible without large databases being stored on computer systems.

New technology is not without problems in its introduction. The US census had intended to use the handheld computers but cost escalated and this was abandoned, with the contract being sold to Brazil. Online response is a good idea but one of the functions of census is to make sure everyone is counted accurately. A system which allowed people to enter their address without verification would be open to abuse. Therefore households have to be verified on the ground, typically by an enumerator visit or post out. Paper forms are still necessary for those without access to Internet connections. It is also plausible that the hidden nature of an administrative census means that users are not engaged with the importance of contributing their data to official statistics.

Uses of Census data

In the nineteenth century, the first censuses collected paper enumerations that had to be collated by hand so the statistical uses were very basic. The government owned the data and were able to publish statistics themselves on the state of the nation. Uses were to measure changes in the population and apportion representation. Population estimates could be compared to those of other countries.

By the beginning of the twentieth century, censuses were recording households and some indications of their employment. In some countries, census archives are released for public examination after many decades, allowing genealogists to track the ancestry of interested people. Archives provide a substantial historical record which may challenge established notions of tradition. It is also possible to understand the societal history through job titles and arrangements for the destitute and sick.

As governments assumed responsibility for schooling and welfare, large government departments made extensive use of census data. Actuarial estimates could be made to project populations and plan for provision in local government and regions. It was also possible for central government to allocate funding on the basis of census data.

Even into the mid twentieth century, census data was only directly accessible to large government departments. However, computers meant that tabulations could be used directly by university researchers, large businesses and local government offices. They could use the detail of the data to answer new questions and add to local and specialist knowledge.

Now, census data are published in a wide variety of formats to be accessible to business, all levels of governance, media, students and teachers, charities and researchers, and any citizen who is interested. Data can be represented visually or analysed in complex statistical models, to show the difference between certain areas, or to understand the association between different personal characteristics. Census data offer a unique insight into small areas and small demographic groups which sample data would be unable to capture with precision.

Privacy

Although the census provides a useful way of obtaining statistical information about a population, such information can sometimes lead to abuses, political or otherwise, made possible by the linking of individuals' identities to anonymous census data.[15] This consideration is particularly important when individuals' census responses are made available in microdata form, but even aggregate-level data can result in privacy breaches when dealing with small areas and/or rare subpopulations.

For instance, when reporting data from a large city, it might be appropriate to give the average income for black males aged between 50 and 60. However, doing this for a town that only has two black males in this age group would be a breach of privacy because either of those persons, knowing his own income and the reported average, could determine the other man's income.

Typically, census data are processed to obscure such individual information. Some agencies do this by intentionally introducing small statistical errors to prevent the identification of individuals in marginal populations;[16] others swap variables for similar respondents. Whatever measures have been taken to reduce the privacy risk in census data, new technology in the form of better electronic analysis of data poses increasing challenges to the protection of sensitive individual information. This known as statistical disclosure control.

Another possibility is to present survey results by means of statistical models in the form of a multivariate distribution mixture.[17] The statistical information in the form of conditional distributions (histograms) can be derived interactively from the estimated mixture model without any further access to the original database. As the final product does not contain any protected microdata, the model based interactive software [18] can be distributed without any confidentiality concerns.

Another method is simply to release no data at all, except very large scale data directly to the central government. Different release strategies between government have led to an international project (IPUMS) to co-ordinate access to microdata and corresponding metadata. Such projects also promote standardising metadata by projects such as SDMX so that best use can be made of the minimal data available.

Historical examples

Egypt

Censuses in Egypt are said to have been taken during the early Pharaonic period in 3340 BCE and in 3056 BCE.

Ancient Greece

There are several accounts of ancient Greek and Mesopotamian city states carrying out censuses.[19] The question of which is first is clouded by very different approaches: counting only men, counting a pile of rocks etc. but such censuses took place 1600 BCE and earlier.

Ancient Israel

Censuses are mentioned in the Bible. God commands a flat tax to be paid with the census in Exodus 30:11-16 for the upkeep of the Tabernacle. The Book of Numbers is named after the counting of the Israelite population (in Numbers 1-4) according to the house of the Fathers after the exodus from Egypt. A second census was taken while the Israelite were camped in the plains of Moab, in Numbers 26.

King David performed a census that produced disastrous results (in Samuel 2 24 and Chronicles 1 21). His son, King Solomon, had all of the foreigners in Israel counted in Chronicles 2 2:17.

The Gospel of Luke records Jesus being born during a census in Luke 2.

China

In 2 CE during the Han Dynasty, China held a census still considered by scholars to be quite accurate.[20] [21] The census is considered one of the world's earliest preserved censuses,[] and found 57.67 million people registered in 12.36 million households.[22] [23] The areas with the highest population densities were the Yellow River and Huai River valleys. Chengdu was the largest city in China, reaching a population of 282,147 people.[] Another census dates to AD 144, when 49.73 million people were recorded living in 9.94 million households.

India

The oldest recorded census in India is thought to have occurred around 300 BCE during the reign of the Emperor Chandragupta Maurya under the leadership of Kautilya or Chanakya.[24]

Rome

The word "census" originated in ancient Rome from the Latin word *censere* ("to estimate"). The census played a crucial role in the administration of the Roman Empire, as it was used to determine taxes. With few interruptions, it was usually carried out every five years.[25] It provided a register of citizens and their property from which their duties and privileges could be listed. It is said to have been instituted by the Roman king Servius Tullius in the 6th century BCE,[26] at which time the number of arms-bearing citizens was counted at around 80,000.[27]

Rashidun and Umayyad Caliphates

In the Middle Ages, the Caliphate began conducting regular censuses soon after its formation, beginning with the one ordered by the second Rashidun caliph, Umar.[28]

Medieval Europe

The Domesday Book was undertaken in 1086 CE by William I of England so that he could properly tax the land he had recently conquered in medieval Europe. In 1183 CE, a census was taken of the crusader Kingdom of Jerusalem, to ascertain the number of men and amount of money that could possibly be raised against an invasion by Saladin, sultan of Egypt and Syria.

Inca Empire

In the 15th century, the Inca Empire had a unique way to record census information. The Incas did not have any written language but recorded information collected during censuses and other numeric information as well as non-numeric data on quipus, strings from llama or alpaca hair or cotton cords with numeric and other values encoded by knots in a base-10 positional system.

Spanish empire

On May 25, 1577, King Philip II of Spain ordered by royal cédula the preparation of a general description of Spain's holdings in the Indies. Instructions and a questionnaire, issued in 1577 by the Office of the Cronista Mayor, were distributed to local officials in the Viceroyalties of New Spain and Peru to direct the gathering of information. The questionnaire, composed of fifty items, was designed to elicit basic information about the nature of the land and the life of its peoples. The replies, known as "relaciones geográficas," were written between 1579 and 1585 and were returned to the Cronista Mayor in Spain by the Council of the Indies.

See also

- IBM and the Holocaust
- *Liber Censuum*
- Race and ethnicity in censuses
- Social research

Notes

[2] United Nations (2008). Principles and Recommendations for Population and Housing Censuses (http://unstats.un.org/unsd/demographic/sources/census/docs/P&R_Rev2.pdf). Statistical Papers: Series M No. 67/Rev.2. p8. ISBN 978-92-1-161505-0.

[3] http://www.unece.org/fileadmin/DAM/stats/publications/CES_2010_Census_Recommendations_English.pdf

[6] Leo Breiman (1994). "The 1991 Census Adjustment: Undercount or Bad Data?", *Statistical Science* **9** (4), pp.458-475.

[7] World Population and Housing Census Programme (2010) Post Enumeration Surveys: Operational guidelines (http://unstats.un.org/unsd/demographic/standmeth/handbooks/Manual_PESen.pdf), United Nations Secretariat, Dept of Economic and Social Affairs, Statistics Division, Tech Report

[8] Benton, P. Trout, Catfish and Roach: The beginner's guide to census population estimates (http://www.ons.gov.uk/ons/guide-method/census/2011/the-2011-census/census-coverage-survey/trout--catfish-and-roach.pdf), Office for National Statistics, UK

[9] Other methods of census taking (http://www.ons.gov.uk/ons/guide-method/census/2011/what-is-a-census/other-methods-of-census-taking/index.html), Office for National Statistics, UK

[15] The Census and Privacy (http://www.epic.org/privacy/census/).

[18] http://ro.utia.cas.cz/census

[21] Twitchett, D., Loewe, M., and Fairbank, J.K. *Cambridge History of China: The Ch'in and Han Empires 221 B.C.-A.D. 220*. Cambridge University Press (1986), p. 240.

[22] Nishijima (1986), 595–596.

[23] H. Yoon (1985). "An early Chinese idea of a dynamic environmental cycle", *GeoJournal* **10** (2), 211-212.

[25] Scheidel, Walter (2009) *Rome and China: comparative perspectives on ancient world empires*. Oxford University Press, p. 28.

[26] Livy *Ab urbe condita* 1.42

[27] Livy *Ab urbe condita* 1.42, citing Fabius Pictor

References

- Alterman, Hyman, (1969). *Counting People: The Census in History*. Harcourt, Brace & Company.
- Bielenstein, Hans, (1978). "Wang Mang, the restoration of the Han dynasty, and Later Han." In *The Cambridge History of China*, vol. 1, eds. Denis Twitchett and John K. Fairbank, p. 223-90, Cambridge: Cambridge University Press.
- Krüger, Stephen, (Fall 1991). "The Decennial Census", 19 *Western State University Law Review* 1 (http://wsulawreview.org/Volume19Articles.pdf); available at HeinOnline (http://heinonline.org/HOL/LandingPage?collection=journals&handle=hein.journals/wsulr19&div=6&id=&page=) (subscription required).
- Effects of UK 'Jedi' hoax on 2001 UK census from ONS (http://www.statistics.gov.uk/cci/nugget.asp?id=297).
- U.S. Census Press Release on 1930 Census (http://www.census.gov/Press-Release/www/releases/archives/miscellaneous/000507.html).
- U.S. Census Press Release on Soundex and WPA (http://www.archives.gov/publications/prologue/spring_2002_soundex_projects.html).

External links

- My Free Census (http://www.myfreecensus.com/) Find your ancestors in the U.S. census, and International census, with these free online census records, census images, census indexes and census databases in our easy to use census directory.
- Census of Ireland 1911 (http://www.census.nationalarchives.ie/).
- Online Historical Population Reports Project (OHPR) (http://www.histpop.org).
- PR as a function of census management: comparative analysis of fifteen census experiences (http://www.tesionline.com/intl/thesis.jsp?idt=8384)

Population_density

Population density (in agriculture **standing stock** and standing crop) is a measurement of population per unit area or unit volume. It is frequently applied to living organisms, and particularly to humans. It is a key geographic term.[1]

Population density (people per km^2) by country, 2006

Biological population densities

Population density is population divided by total land area or water volume, as appropriate.[1]

Low densities may cause an extinction vortex and lead to further reduced fertility. This is called the Allee effect after the scientist who identified it. Examples of the causes in low population densities include:[2]

- Increased problems with locating sexual mates
- Increased inbreeding

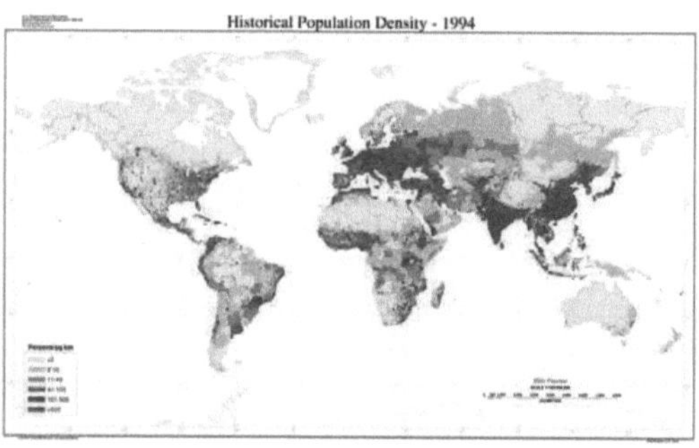

Population density (people per km^2) map of the world in 1994 (detailed).

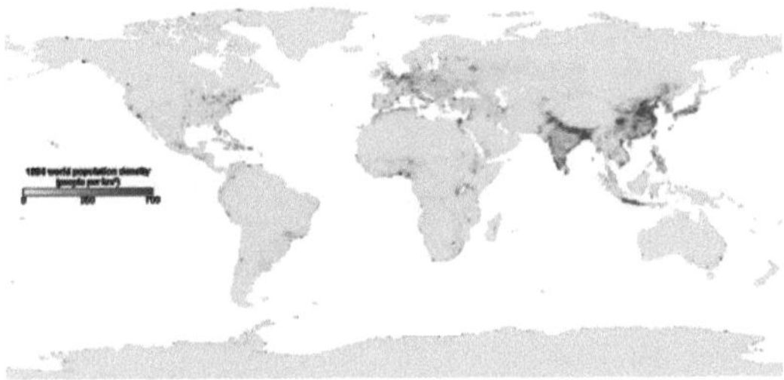

Population density (people per km^2) map of the world in 1994.

Deserts around the world. Compare with maps above.

Human population density

Monaco in South Europe, currently holds the record for being the most densely populated nation in the world.

For humans, population density is the number of people per unit of area usually per square kilometer or mile (which may include or exclude cultivated or potentially productive area). Commonly this may be calculated for a county, city, country, another territory, or the entire world.

The world's population is 7 billion,[3] and Earth's total area (including land and water) is 510 million square kilometers (197 million square miles).[4] Therefore the worldwide human population density is 6.8 billion ÷ 510 million = 13.3 per km^2 (34.5 per sq. mile). If only the Earth's land area of 150 million km^2 (58 million sq. miles) is taken into account, then human population density increases to 45.3 per km^2 (117.2 per sq. mile). This calculation includes all continental and island land area, including Antarctica. If Antarctica is also excluded, then population density rises to 50 people per km^2 (129.28 per sq. mile).[1] Considering that over half of the Earth's land mass consists of areas inhospitable to human inhabitation, such as deserts and high mountains, and that population tends to cluster around seaports and fresh water sources, this number by itself does not give any meaningful measurement of human population density.

Several of the most densely populated territories in the world are city-states, microstates, or dependencies.[] [5] These territories share a relatively small area and a high urbanization level, with an economically specialized city population drawing also on rural resources outside the area, illustrating the difference between high population density and overpopulation.

Mongolia is the least densely populated country in the world.

Cities with high population densities are, by some, considered to be overpopulated, though the extent to which this is the case depends on factors like quality of housing and infrastructure and access to resources.[6] Most of the most densely populated cities are in southern and eastern Asia, though Cairo and Lagos in Africa also fall into this category.[7]

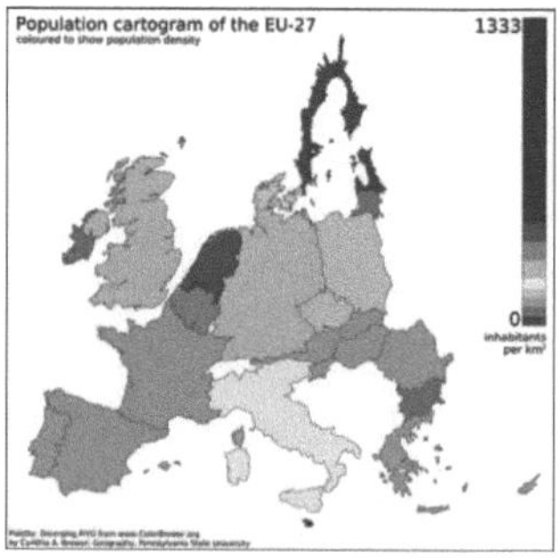

A population cartogram uses areas rather than colors to represent population.

City population is, however, heavily dependent on the definition of "urban area" used: densities are often higher for the central municipality itself, than when more recently developed and administratively separate suburban communities are included, as in the concepts of agglomeration or metropolitan area, the latter including sometimes neighboring cities. For instance, Milwaukee has a greater population density when just the inner city is measured, and not the surrounding suburbs as well.[8]

As a comparison, based on a world population of seven billion, the world's inhabitants would, as a loose crowd taking up ten square feet (one square metre) per person (Jacobs Method), would occupy a space a little larger than Delaware's land area.

Most densely populated countries

With population above 1 million

Rank	Country/Region	Population	Area (km^2)	Density (Pop. per km^2)
1	Singapore	5,183,700	710	7301
2	Hong Kong	7,061,200	1,104	6396
3	Bahrain	1,234,596	750	1646
4	Bangladesh	152,518,015	147,570	1034
5	Taiwan (R.O.C)	22,955,395	36,190	634
6	Mauritius	1,288,000	2,040	631
7	South Korea	48,456,369	99,538	487
8	Rwanda	10,718,379	26,338	407
9	Netherlands	16,760,000	41,526	404
10	Lebanon	4,224,000	10,452	404

With population above 10 million

Rank	Country/Region	Population	Area (km^2)	Density (Pop. per km^2)
1	Bangladesh	152,518,015	147,570	1034
2	Taiwan (R.O.C)	22,955,395	36,190	634
3	South Korea	48,456,369	99,538	487
4	Rwanda	10,718,379	26,338	407
5	Netherlands	16,760,000	41,526	404
6	India	1,210,193,422	3,185,263	380
7	Belgium	11,007,020	30,528	361
8	Japan	127,960,000	377,944	339
9	Sri Lanka	20,653,000	65,610	345

Other methods of measurement

While arithmetic density is the most common way of measuring population density, several other methods have been developed which aim to provide a more accurate measure of population density over a specific area.

- **Arithmetic density**: The total number of people / area of land (measured in km^2 or sq miles).
- **Physiological density**: The total population / area of arable land.
- **Agricultural density**: The total *rural* population / area of arable land.
- **Residential density** : The number of people living in an urban area / area of residential land.
- **Urban density** : The number of people inhabiting an urban area / total area of urban land.
- **Ecological optimum**: The density of population which can be supported by the natural resources.

See also

- Demography
- Human geography
- Idealized population
- Optimum population
- Population bottleneck
- Population genetics
- Population health
- Population momentum
- Population pyramid
- Rural transport problem
- Small population size
- Distance sampling

Lists:

- List of cities by population
- List of cities by population density
- List of European cities proper by population density
- List of islands by population density
- List of countries by population density
- List of U.S. states by population density

References

[1] Matt Rosenberg Population Density (http://geography.about.com/od/populationgeography/a/popdensity.htm). Geography.about.com. March 2 2011. Retrieved on 2011-12-10.
[2] Minimum viable population size (http://www.eoearth.org/article/Minimum_viable_population_size). Eoearth.org (2010-03-06). Retrieved on 2011-12-10.
[3] U.S. & World Population Clocks (http://www.census.gov/main/www/popclock.html). Census.gov. Retrieved on 2011-12-10.
[4] World (https://www.cia.gov/library/publications/the-world-factbook/geos/xx.html). CIA World Handbook
[5] The Monaco government uses a smaller surface area figure resulting in a population density of 18,078 per km^2
[6] Human Population (http://www.globalissues.org/issue/198/human-population). Global Issues. Retrieved on 2011-12-10.
[7] The largest cities in the world by land area, population and density (http://www.citymayors.com/statistics/largest-cities-density-125.html). Citymayors.com. Retrieved on 2011-12-10.
[8] The Population of Milwaukee County (http://www.wisconline.com/greenmap/milwaukee/population.html). Wisconline.com. Retrieved on 2011-12-10.

External links

- Selected Current and Historic City, Ward & Neighborhood Densities (http://www.demographia.com/db-citydenshist.htm)

Race_and_ethnicity_in_the_United_States_Census

Race and ethnicity in the United States Census, defined by the federal Office of Management and Budget (OMB) and the United States Census Bureau, are self-identification data items in which residents choose the race or races with which they most closely identify, and indicate whether or not they are of Hispanic Latino origin (the only categories for ethnicity).[1] [2]

The racial categories represent a social-political construct for the race or races that respondents consider themselves to be and "generally reflect a social definition of race recognized in this country."[] OMB defines the concept of race as outlined for the US Census as not "scientific or anthropological" and takes into account "social and cultural characteristics as well as ancestry", using "appropriate scientific methodologies" that are not "primarily biological or genetic in reference."[3] The race categories include both racial and national-origin groups.[]

Race and ethnicity are considered separate and distinct identities, with Hispanic or Latino origin asked as a separate question. Thus, in addition to their race or races, all respondents are categorized by membership in one of two ethnic categories, which are "Hispanic or Latino" and "Not Hispanic or Latino". However, the practice of separating "race" and "ethnicity" as different categories has been criticized both by the American Anthropological Association and members of U.S. Commission on Civil Rights.[4] [5]

In 1997, OMB issued a Federal Register Notice regarding revisions to the standards for the classification of federal data on race and ethnicity.[6] OMB developed race and ethnic standards in order to provide "consistent data on race and ethnicity throughout the Federal Government. The development of the data standards stem in large measure from new responsibilities to enforce civil rights laws." Among the changes, OMB issued the instruction to "mark one or more races" after noting evidence of increasing numbers of interracial children and wanting to capture the diversity in a measurable way and having received requests by people who wanted to be able to acknowledge their or their children's full ancestry rather than identifying with only one group. Prior to this decision, the Census and other government data collections asked people to report only one race.[]

How data on race and ethnicity are used

The OMB states that "many federal programs are put into effect based on the race data obtained from the decennial census (i.e., promoting equal employment opportunities; assessing racial disparities in health and environmental risks). Race data are also critical for the basic research behind many policy decisions. States require these data to meet legislative redistricting requirements. The data are needed to monitor compliance with the Voting Rights Act by local jurisdictions".

"Data on ethnic groups are important for putting into effect a number of federal statutes (i.e., enforcing bilingual election rules under the Voting Rights Act; monitoring and enforcing equal employment opportunities under the Civil Rights Act). Data on Ethnic Groups are also needed by local governments to run programs and meet legislative requirements (i.e., identifying segments of the population who may not be receiving medical services under the Public Health Act; evaluating whether financial institutions are meeting the credit needs of minority populations under the Community Reinvestment Act)."[]

Brief overview of race and ethnicity in the U.S. Census' history

Further information: Race and ethnicity in the United States

Further information: Race and ethnicity in censuses#United States

Census 1790

In 1790, first official year of the U.S. Census, the following questions were asked, four of which had racial implications:

- Number of free White males aged under 16 years
- Number of free White males aged 16 years and upward
- Number of free White females
- Number of other free persons
- Number of slaves[]

In 1800 and 1810, the age question regarding free white males was more detailed.[]

Census 1820

The 1820 census built on the questions asked in 1810 by asking age questions about the slaves who were formerly owned. Also the term "colored" enters the census nomenclature. In addition, a question stating "Number of foreigners not naturalized" was included.[]

Census 1830

For the 1830 census, a new question which stated "The number of White persons who were foreigners not naturalized" was included.[] This reflected the growth of Nativist movements in American society at this time - as well as combining the number and age question of both slaves and free colored individuals.

Census 1850

The 1850 census saw a dramatic shift in the way information about residents was collected. For the first time, free persons were listed individually instead of by head of household. There were two questionnaires: one for free inhabitants and one for slaves. The question on the free inhabitants schedule about color was a column that was to be left blank if a person was white, marked "B" if a person was black, and marked "M" if a person was mulatto. Slaves were listed by owner, and classified by gender and age, not individually, and the question about color was a column that was to be marked with a "B" if the slave was black and an "M" if mulatto.[]

Census 1870

For the 1870 census, the color/racial question was expanded to include "C" for Chinese, which was a category that included all east Asians, as well as "I" for American Indians.[]

Census 1890

For 1890, the Census Office changed the design of the population questionnaire. Residents were still listed individually, but a new questionnaire sheet was used for each family. Additionally, this was the first year that the census distinguished between different East Asian races, such as Japanese and Chinese, due to increased immigration. This census also marked the beginning of the term "race" in the questionnaires. Enumerators were instructed to write "White," "Black," "Mulatto," "Quadroon," "Octoroon," "Chinese," "Japanese," or "Indian."[]

Census 1900

For 1900, the "Color or Race" question was slightly modified, removing the term "Mulatto". Also, there was an inclusion of an "Indian Population Schedule" in which "enumerators were instructed to use a special expanded questionnaire for American Indians living on reservations or in family groups off of reservations." This expanded version included the question "Fraction of person's lineage that is white."[]

Census 1910

The 1910 census was similar to that of 1900, but it included a re-insertion of "Mulatto" and a question about the respondent's "mother tongue." "Ot" was also added to signify "other races", with space for a race to be written in. This decade's version of the Indian Population Schedule featured questions asking the individual's proportion of white, black, or American Indian lineage.[]

Census 1920

The 1920 census questionnaire was similar to 1910, but excluded a separate schedule for American Indians. "Hin", "Kor", and "Fil" were also added to the "Color or Race" question, signifying Hindu (South Asia Indian), Korean, and Filipino, respectively.[]

Census 1930

The biggest change in this year's census was in racial classification. Enumerators were instructed to no longer use the "Mulatto" classification. Instead, they were given special instructions for reporting the race of interracial persons. A person with both white and black ancestry (termed "blood") was to be recorded as "Negro," no matter the fraction of that lineage (the "one-drop rule"). A person of mixed Black and American Indian ancestry was also to be recorded as "Neg" (for "Negro") unless he was considered to be "predominantly" American Indian and accepted as such within the community. A person with both White and American Indian ancestry was to be recorded as an Indian, unless his American Indian ancestry was small, and he was accepted as White within the community. In all situations in which a person had White and some other racial ancestry, he was to be reported as that other race. Persons who had minority interracial ancestry were to be reported as the race of their father.

For the first and only time, "Mexican" was listed as a race. Enumerators were instructed that all persons born in Mexico, or whose parents were born in Mexico, should be listed as Mexicans, and not under any other racial category. But, in prior censuses and in 1940, enumerators were instructed to list Mexican Americans as white.[7]

The Supplemental American Indian questionnaire was back, but in abbreviated form. It featured a question asking if the person was of full or mixed American Indian ancestry.[] [8]

Census 1940

President Franklin D. Roosevelt promoted a "good neighbor" policy that sought better relations with Mexico. In 1935 a federal judge ruled that three Mexican immigrants were ineligible for citizenship because they were not white, as required by federal law. Mexico protested, and Roosevelt decided to circumvent the decision and make sure the federal government treated Hispanics as white. The State Department, the Census Bureau, the Labor Department, and other government agencies therefore made sure to uniformly classify people of Mexican descent as white. This policy encouraged the League of United Latin American Citizens in its quest to minimize discrimination by asserting their whiteness.[9]

The 1940 census was the first to include separate population and housing questionnaires.[] The race category of "Mexican" was eliminated in 1940, and the population of Mexican descent was counted with the White population.[7]

Census 1950

The 1950 Census questionnaire removed the word “color” from the racial question, and also removed Hindu and Korean from the race choices.[]

Census 1960

The 1960 Census re-added the word “color” to the racial question, and changed “Indian” to “American Indian”, as well as added Hawaiian, Part-Hawaiian, Aleut, and Eskimo. The Other (print out race) option was removed.[]

Census 1970

This year's census included “Negro or Black”, re-added Korean and the Other race option. East Indians (the term used at that time for people whose ancestry is from the Indian subcontinent) were counted as White. There was a questionnaire that was asked of only a sample of respondents. These questions were as follows:

a. Where was this person born?

b. Is this person's origin or descent...

- Mexican
- Puerto Rican
- Cuban
- Central or South American
- Other Spanish
- None of These

14. What country was the person's father born in?

15. What country was the person's mother born in?

16.

a. For persons born in a foreign country- Is the person naturalized?

b. When did the person come to the United States to stay?

17. What language, other than English, was spoken in the person's home as a child?

- Spanish
- French
- German
- Other
- None, only English[]

Census 1980

This year added several options to the race question, including Vietnamese, Indian (East) Guamanian, Samoan, and re-added Aleut. Again, the term "color" was removed from the racial question, and the following questions were asked of a sample of respondents:

11. In what state or foreign country was the person born?

12. If this person was born in a foreign country...

a. Is this person a naturalized citizen of the United States?

b. When did this person come the United States to stay?

13.

a. Does this person speak a language other than English at home?

b. If yes, what is this language?

c. If yes, how well does this person speak English?

14. What is this person's ancestry?[]

Census 1990

The racial categories in this year are as they appear in the 2000 and 2010 Census. The following questions were asked of a sample of respondents for the 1990 Census:

8. In what U.S. State or foreign country was this person born?

9. Is this person a citizen of the United States?

10. If this person was not born in the United States, when did this person come to the United States to stay?[]

The 1990 Census was not designed to capture multiple racial responses, and when individuals marked the Other race option and provided a multiple write in, the response was assigned according to the race written first. "For example, a write in of "Black-White" was assigned a code of Black, a write in of "White-Black" was assigned a code of White."[]

> *In the United States, census data indicate that the number of children in interracial families grew from less than one half million in 1970 to about two million in 1990. In 1990, for interracial families with one white American partner, the other parent...was Asian American for 45 percent...*[10]

Census 2000

Race was asked differently in the Census 2000 in several other ways than previously. Most significantly, respondents were given the option of selecting one or more race categories to indicate racial identities. Data show that nearly seven million Americans identified as members of two or more races. Because of these changes, the Census 2000 data on race are not directly comparable with data from the 1990 census or earlier censuses. Use of caution is therefore recommended when interpreting changes in the racial composition of the US population over time.

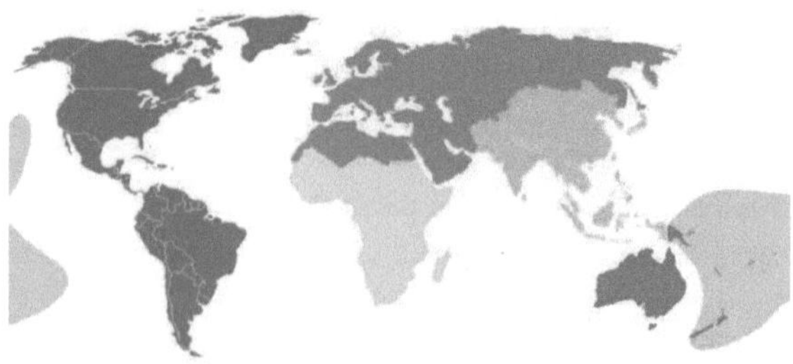

The race of people based on their or their ancestral country of origin. This is the country race categorization that was used for the 2000 U.S. Census.

The following definitions apply to the 2000 census only.[]

- "White. A person having origins in any of the original peoples of Europe, the Middle East, or North Africa. It includes people who indicate their race as "White" or report entries such as Irish, German, Scottish, Italian, Near Easterner, Arab, or Polish.[]
- "Black or African American. A person having origins in any of the Black racial groups of Africa. It includes people who indicate their race as 'Black, African Am.' or provide written entries such as Kenyan, Nigerian, or Haitian."[]
- "American Indian and Alaska Native. A person having origins in any of the original peoples of North and South America (including Central America) and who maintain tribal affiliation or community attachment."[]
- "Asian. A person having origins in any of the original peoples of the Far East, Southeast Asia, or the Indian subcontinent including, for example, Cambodia, China, India, Japan, Korea, Malaysia, Pakistan, the Philippine Islands, Thailand, and Vietnam. It includes 'Asian Indian,' 'Chinese', 'Filipino', 'Korean', 'Japanese', 'Vietnamese', and 'Other Asian'."[]
- "Native Hawaiian and Other Pacific Islander. A person having origins in any of the original peoples of Hawaii, Guam, Samoa, or other Pacific Islands. It includes people who indicate their race as 'Native Hawaiian', 'Guamanian or Chamorro', 'Samoan', and 'Other Pacific Islander'."[]
- "Some other race. Includes all other responses not included in the 'White', 'Black or African American', 'American Indian and Alaska Native', 'Asian' and 'Native Hawaiian and Other Pacific Islander' race categories described above. Respondents providing write-in entries such as multiracial, mixed, interracial, We-Sort, or a Hispanic/Latino group (for example, Mexican, Puerto Rican, or Cuban) in the "Some other race" category are included here."[]
- "Two or more races. People may have chosen to provide two or more races either by checking two or more race response check boxes, by providing multiple write-in responses, or by some combination of check boxes and write-in responses."[]

Snapshot: Race in the US Census
The 23rd federal census, 2010[11] asks one ethnic and one race question (questions 1-4 not reproduced here, questions 5 and 6 paraphrased): 8. Is the person of Hispanic, Latino, or Spanish origin? • No, not of Hispanic, Latino, or Spanish origin • Yes, Mexican, Mexican Am., Chicano • Yes, Puerto Rican • Yes, Cuban • Yes, another Hispanic, Latino, or Spanish origin — Print origin, for example, Argentinean, Colombian, Dominican, Nicaraguan, Salvadoran, Spaniard, and so on.

9. What is the person's race?

- White
- Black or African American
- American Indian or Alaska Native — Print name of enrolled or principal tribe.
- Asian Indian
- Chinese
- Filipino
- Other Asian — Print race, for example, Hmong, Laotian, Thai, Pakistani, Cambodian, and so on.
- Japanese
- Korean
- Vietnamese
- Native Hawaiian
- Guamanian or Chamorro
- Samoan
- Other Pacific Islander — Print race, for example, Fijian, Tongan, and so on.
- Some other race — Print race.

This census acknowledged that "race categories include both racial and national-origin groups."

The federal government of the United States has mandated that "in data collection and presentation, federal agencies are required to use a minimum of two ethnicities: 'Hispanic or Latino' and 'Not Hispanic or Latino'."[] The Census Bureau defines "Hispanic or Latino" as "a person of Cuban, Mexican, Puerto Rican, South or Central American or other Spanish culture or origin regardless of race."[] For discussion of the meaning and scope of the Hispanic or Latino ethnicity, see the Hispanic and Latino Americans and Racial and ethnic demographics of the United States articles.

Use of the word *ethnicity* for Hispanics only is considerably more restricted than its conventional meaning, which covers other distinctions, some of which are covered by the "race" and "ancestry" questions. The distinct questions accommodate the possibility of Hispanic and Latino Americans' also declaring various racial identities (see also White Hispanic and Latino Americans, Asian Latinos, and Black Hispanic and Latino Americans).

In the 2000 Census, 12.5% of the US population reported "Hispanic or Latino" ethnicity and 87.5% reported "Not-Hispanic or Latino" ethnicity.[]

Census 2010

The 2010 US Census included changes designed to more clearly distinguish Hispanic ethnicity as not being a race. That included adding the sentence: "For this census, Hispanic origins are not races."[12] [] Additionally, the Hispanic terms were modified from "Hispanic or Latino" to "Hispanic, Latino or Spanish origin".[12] []

Although used in the Census and the American Community Survey, "Some other race" is not an official race,[] and the Bureau considered eliminating it prior to the 2000 Census.[] As the 2010 census form did not contain the question titled "Ancestry" found in prior censuses, there were campaigns to get non-Hispanic West Indian Americans,[13] Turkish Americans,[14] Arab Americans and Iranian Americans to indicate their ethnic or national background through the race question, specifically the "Some other race" category.[15] [16]

The Interagency Committee has suggested that the concept of marking multiple boxes be extended to the Hispanic origin question, thereby freeing individuals from having to choose between their parents' ethnic heritages. In other words, a respondent could choose both "Hispanic or Latino" and "Not Hispanic or Latino".[]

Relation between ethnicity and race in census results

The Census Bureau warns that data on race in 2000 Census are not directly comparable to those collected in previous censuses.[] Many residents of the United States consider race and ethnicity to be the same.[3]

Population distribution by race (2000 census)[17]

Race	Hispanic or Latino	% of H/L	% of US	Not Hispanic or Latino	% of Not H/L	% of US
Any races	35,305,818	100	12.5	246,116,088	100	87.5
One race:	33,081,736	93.7	11.8	241,513,942	98.1	85.8
White	16,907,852	47.9	6.0	194,552,774	79.1	69.1
Black or African A.	710,353	2.0	0.3	33,947,837	13.8	12.1
A. Indian/ Alaska Nat.	407,073	1.2	0.1	2,068,883	0.8	0.7
Asian	119,829	0.3	<0.1	10,123,169	4.1	3.6
Hawaiian N. & Pacific Is.	45,326	0.1	<0.1	353,509	0.1	0.1
Some other	14,891,303	42.2	5.3	467,770	0.2	0.2
2+ races:	2,224,082	6.3	0.8	4,602,146	1.9	1.6
Some other + W/B/N/A	1,859,538	5.3	0.7	1,302,875	0.5	0.5
2+ W/B/N/A	364,544	1.0	0.1	3,299,271	1.3	1.2

In the 2000 Census, respondents were tallied in each of the race groups they reported. Consequently, the total of each racial category exceeds the total population because some people reported more than one race.[]

According to James P. Allen and Eugene Turner from California State University, Northridge, by some calculations in the 2000 Census the largest part white bi-racial population is white/Native American and Alaskan Native, at 7,015,017, followed by white/black at 737,492, then white/Asian at 727,197, and finally white/Native Hawaiian and Other Pacific Islander at 125,628.[18]

The Census Bureau implemented a Census Quality Survey, gathering data from approximately 50,000 households in order to assess the reporting of race and Hispanic origin in the 2000 Census with the purpose creating a way to make comparisons between the 2000 Census with previous Census racial data.[]

In September 1997, during the process of revision of racial categories previously declared by OMB directive no. 15, the American Anthropological Association (AAA) recommended that OMB combine the "race" and "ethnicity" categories into one question to appear as "race/ethnicity" for the 2000 US Census. The Interagency Committee agreed, stating that "race" and "ethnicity" were not sufficiently defined and "that many respondents conceptualize 'race' and 'ethnicity' as one in the same [*sic*] underscor[ing] the need to consolidate these terms into one category, using a term that is more meaningful to the American people."[3]

The AAA also stated,

> "The American Anthropological Association recommends the elimination of the term 'race' from OMB Directive 15 during the planning for the 2010 Census. During the past 50 years, 'race' has been scientifically proven to not be a real, natural phenomenon. More specific, social categories such as 'ethnicity' or 'ethnic group' are more salient for scientific purposes and have fewer of the negative, racist connotations for which the concept of race was developed."

> "Yet the concept of race has become thoroughly—and perniciously—woven into the cultural and political fabric of the United States. It has become an essential element of both individual identity and government policy. Because so much harm has been based on 'racial' distinctions over the years, correctives for such harm must also acknowledge the impact of 'racial' consciousness among the U.S. populace, regardless of the fact that 'race' has no scientific justification in human biology. Eventually, however, these classifications must be transcended and replaced by more non-racist and accurate ways of representing the diversity of the U.S. population."[3]

The recommendations of the AAA were not adopted by the Census Bureau for the 2000 Census or the 2010 Census.

Other agencies

In 2001, the National Institutes of Health adopted the new language to comply with the revisions to Directive 15,[19] as did the Equal Employment Opportunity Commission of the United States Department of Labor in 2007.[20] See Race and ethnicity (EEO).

See also

- Historical racial and ethnic demographics of the United States
- Language (United States Census)
- Race (classification of humans)
- Race and ethnicity in censuses
- Race and ethnicity in the United States

References

[3] American Anthropological Association. "A Brief History of the OMB Directive 15." 1997. May 18, 2007. (http://www.aaanet.org/gvt/ombdraft.htm)

[7] The 1930 Census in Perspective (http://www.1930census.com/us_census_history.php), 1930census.com

[8] For an image copy of the full 1930 census instructions, see http://www.lib.ncsu.edu/guides/popcensus/images/1930enu.pdf.

[9] Patrick D. Lukens, *A Quiet Victory for Latino Rights: FDR and the Controversy over Whiteness* (University of Arizona Press, 2012)

[10] U.S. Census Bureau, 2000 (http://www2.census.gov/census_2000/datasets/CQS/B.3.pdf)

[12] Waite, Preston. US Census Bureau. " 2010 Decennial Census Program (http://www.census.gov/sdc/www/0306waite.ppt)." 2006. accessed July 7, 2008.

[13] Archive copy (http://web.archive.org/web/*/http://hosted.ap.org/dynamic/stories/U/US_CENSUS_CARIBBEAN_IDENTITY?SITE=NYMID&SECTION=HOME&TEMPLATE=DEFAULT) at the Wayback Machine

[15] Archive copy (http://web.archive.org/web/*/http://www.aaiusa.org/get-involved/4470) at the Wayback Machine

[17] (see Table 1. Population by Hispanic or Latino Origin and by Race for the United States: 2000 and 2010)

[18] http://www.csupomona.edu/~mreibel/2000_Census_Files/Allen-Turner.doc

[20] Final Revisions of the Employer Information Report (EEO-1) (http://www.eeoc.gov/eeo1/index.html) by the EEOC. The page contains links to FAQs, forms and instructions

Per_capita_income

Per capita income, also known as income per person, is the mean income of the people in an economic unit such as a country or city. It is calculated by taking a measure of all sources of income in the aggregate (such as GDP or Gross national income) and dividing it by the total population.

Per capita income as a measure of prosperity

Per capita income is often used as average income, a measure of the wealth of the population of a nation, particularly in comparison to other nations. Per capita income is often used to measure a country's standard of living. It is usually expressed in terms of a commonly used international currency such as the Euro or United States dollar, and is useful because it is widely known, easily calculated from readily-available GDP and population estimates, and produces a useful statistic for comparison of wealth between sovereign territories.this helps the country to know their development status.

Critics claim that per capita income has several weaknesses as an accurate measurement of prosperity:

- Comparisons of per capita income over time need to take into account changes in prices. Without using measures of income adjusted for inflation, they will tend to overstate the effects of economic growth.
- International comparisons can be distorted by differences in the costs of living between countries that aren't reflected in exchange rates. Where the objective of the comparison is to look at differences in living standards between countries, using a measure of per capita income adjusted for differences in purchasing power parity more accurately reflects the differences in what people are actually able to buy with their money.
- As it is a mean value, it does not reflect income distribution. If the distribution of income within a country is skewed, a small wealthy class can increase per capita income far above that of the majority of the population. In this respect Median income is a more useful measure of prosperity than per capita income, because it is less influenced by the outliers.
- Economic activity that does not result in monetary income, such as services provided within the family, or for barter, is usually not counted. The importance of these services varies widely among different economies.

See also

- List of countries by GDP (nominal) per capita—GDP at market or government official exchange rates per inhabitant
- List of countries by GDP (PPP) per capita—GDP calculated at purchasing power parity (PPP) exchange per inhabitant
- List of countries by GNI (nominal, Atlas method) per capita
- List of countries by GNI (PPP) per capita
- List of countries by income equality
- Total personal income

Poverty_threshold

The **poverty threshold**, or **poverty line**, is the minimum level of income deemed adequate in a given country.[1] In practice, like the definition of poverty, the official or common understanding of the poverty line is significantly higher in developed countries than in developing countries.[2] [3] The common **international poverty line** has in the past been roughly $1 a day.[4] In 2008, the World Bank came out with a revised figure of $1.25 at 2005 purchasing-power parity (PPP).[5]

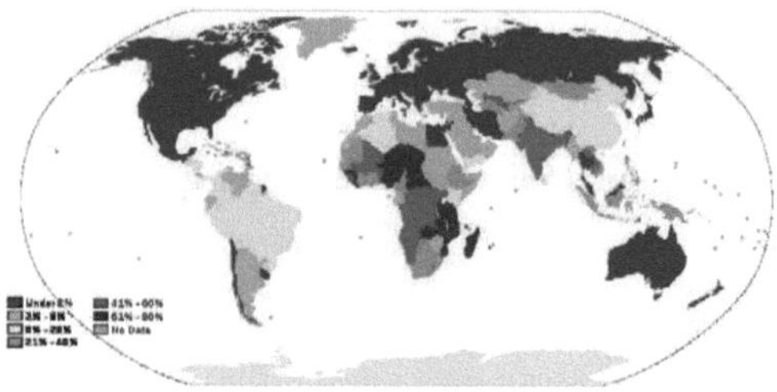

Percentage of population living on less than $1.25 per day. UN estimates 2000-2007.

Determining the poverty line is usually done by finding the total cost of all the essential resources that an average human adult consumes in one year.[6] The largest of these expenses is typically the rent required to live in an apartment, so historically, economists have paid particular attention to the real estate market and housing prices as a strong poverty line affector. Individual factors are often used to account for various circumstances, such as whether one is a parent, elderly, a child, married, etc. The poverty threshold may be adjusted annually.

History

Charles Booth, a pioneering investigator of poverty in London at the turn of the 20th century, popularised the idea of a *poverty line*, a concept originally conceived by the London School Board.[7] Booth set the line at 10 to 20 shillings, which he considered to be the minimum amount necessary for a family of 4 or 5 people to subsist.[8] Benjamin Seebohm Rowntree (1871–1954), a British sociological researcher, social reformer and industrialist, surveyed rich families in York and drew a poverty line in terms of a minimum weekly sum of money "necessary to enable families ... to secure the necessaries of a healthy life", which included fuel and light, rent, food, clothing, and household and personal items. Based on data from leading nutritionists of the period he calculated the cheapest price for the minimum calorific intake and nutritional balance necessary before people got ill or lost weight. He considered this amount to set his poverty line and concluded that 27.84 percent of the total population of York lived below this poverty line.[9] This result corresponded with that from Charles Booth's study of poverty in London and so challenged the view, commonly held at the time, that abject poverty was a problem particular to London and was not widespread in the rest of Britain. Rowntree distinguished between primary poverty, those who lacking in income and secondary poverty, those who had enough income but spent it elsewhere (1901:295-6).[9]

Absolute poverty

Absolute poverty is a level of poverty as defined in terms of the minimal requirements necessary to afford minimal standards of food, clothing, health care and shelter.[10] For the measure to be absolute, the line must be the same in different countries, cultures, and technological levels. Such an absolute measure should look only at the individual's power to consume and it should be independent of any changes in income distribution. The intuition behind an absolute measure is that mere survival takes essentially the same amount of resources across the world and that everybody should be subject to the same standards if meaningful comparisons of policies and progress are to be made. Notice that if everyone's real income in an economy increases, and the income distribution does not change, absolute poverty will decline.

Measuring poverty by an absolute threshold has the advantage of applying the same standard across different locations and time periods: it makes comparisons easier. On the other hand, it suffers from the disadvantage that any absolute poverty threshold is to some extent arbitrary; the amount of wealth required for survival is not the same in all places and time periods. For example, a person living in far northern Scandinavia requires a source of heat during colder months, while a person living on a tropical island does not.

This type of measure is often contrasted with measures of relative poverty (see below), which classify individuals or families as "poor" not by comparing them to a fixed cutoff point, but by comparing them to others in the population under study.

The term *absolute poverty* is also sometimes used as a synonym for extreme poverty. Absolute poverty is the absence of enough resources (such as money) to secure basic life necessities.

According to a UN declaration that resulted from the World Summit on Social Development in Copenhagen in 1995, absolute poverty is "a condition characterised by severe deprivation of basic human needs, including food, safe drinking water, sanitation facilities, health, shelter, education and information. It depends not only on income but also on access to services."[]

David Gordon's paper, "Indicators of Poverty & Hunger", for the United Nations, further defines absolute poverty as the absence of any two of the following eight basic needs:[]

- *Food:* Body Mass Index must be above 16.
- *Safe drinking water:* Water must not come solely from rivers and ponds, and must be available nearby (less than 15 minutes' walk each way).
- *Sanitation facilities:* Toilets or latrines must be accessible in or near the home.
- *Health:* Treatment must be received for serious illnesses and pregnancy.
- *Shelter:* Homes must have fewer than four people living in each room. Floors must not be made of dirt, mud, or clay.
- *Education:* Everyone must attend school or otherwise learn to read.
- *Information:* Everyone must have access to newspapers, radios, televisions, computers, or telephones at home.
- *Access to services:* This item is undefined by Gordon, but normally is used to indicate the complete panoply of education, health, legal, social, and financial (credit) services.

For example, a person who lives in a home with a mud floor is considered severely deprived of shelter. A person who never attended school and cannot read is considered severely deprived of education. A person who has no newspaper, radio, television, or telephone is considered severely deprived of information. All people who meet any two of these conditions — for example, they live in homes with mud floors *and* cannot read — are considered to be living in absolute poverty.

Relative poverty

Relative poverty is the "most useful measure for ascertaining poverty rates in wealthy developed nations".[] [11] [] [12] [] Relative poverty measure is used by the United Nations Development Program (UNDP), the United Nations Children's Fund (UNICEF), the Organisation for Economic Co-operation and Development (OECD) and Canadian poverty researchers.[] [11] [] [12] [] In the European Union, the "relative poverty measure is the most prominent and most–quoted of the EU social inclusion indicators."[]

"[R]elative poverty reflects better the cost of social inclusion and equality of opportunity in a specific time and space."[13]

"Once economic development has progressed beyond a certain minimum level, the rub of the poverty problem – from the point of view of both the poor individual and of the societies in which they live – is not so much the effects of poverty in any absolute form but the effects of the contrast, daily perceived, between the lives of the poor and the lives of those around them. For practical purposes, the problem of poverty in the industrialized nations today is a problem of relative poverty (page 9)."[14] [13]

History of the concept of relative poverty

In 1776 Adam Smith in the Wealth of Nations argued that poverty is the inability to afford, "not only the commodities which are indispensably necessary for the support of life, but whatever the custom of the country renders it indecent for creditable people, even of the lowest order, to be without."[] [15]

In 1958 J. K. Galbraith argued that, "People are poverty stricken when their income, even if adequate for survival, falls markedly behind that of their community."[] [15]

In 1964 in a joint committee economic President's report in the United States, Republicans endorsed the concept of relative poverty. "No objective definition of poverty exists.... The definition varies from place to place and time to time. In America as our standard of living rises, so does our idea of what is substandard."[16] [15]

In 1965 Rose Friedman argued for the use of relative poverty claiming that the definition of poverty changes with general living standards. Those labelled as poor in 1995, would have had "a higher standard of living than many labelled not poor" in 1965.[17] [15]

In 1979, British sociologist, Peter Townsend published his famous definition, "individuals [...] can be said to be in poverty when they lack the resources to obtain the types of diet, participate in the activities and have the living conditions and amenities which are customary, or are at least widely encouraged or approved, in the societies to which they belong (page 31)."[]

Brian Nolan and Christopher T. Whelan of the Economic and Social Research Institute (ESRI) in Ireland explained that, "[P]overty has to be seen in terms of the standard of living of the society in question."[]

Relative poverty measures are used as official poverty rates by the European Union, UNICEF and the OEDC. The main poverty line used in the OECD and the European Union is based on "economic distance", a level of income set at 60% of the median household income.[18]

Relative poverty versus inequality

A measure of *relative poverty* defines "poverty" as being below some relative poverty threshold. For example, the statement that "those individuals who are employed and whose household equivalised disposable income is below 60% of national median equivalised income are poor" uses a relative measure to define poverty.[] Using this definition, if everyone's real income in an economy increases, but the income distribution stays the same, then the rate of relative poverty will also stay the same.

This means, by its very nature, that there will always be a family living in (relative) poverty, even if they have a very high living standard, unless everyone has almost exactly the same income.

Relative poverty measurements can sometimes produce odd results, especially in small populations. For example, if the median household in a wealthy neighborhood earns US$1 million each year, then a family that earns US$100,000 would be considered poor on the relative poverty scale, even though such a family could meet all of its basic needs and much more. At the other end of the scale, if the median household in a very poor neighborhood earned only 50% of what it needs to buy food, then a person who earned the median income would not be considered poor on a relative poverty scale, even though the person is clearly poor on an absolute poverty scale.

Measures of relative poverty are almost the same as measuring income inequality: If a society gets a more equal income distribution, relative poverty will fall. They point out that if society changed in a way that hurt high earners more than low ones, then relative poverty would decrease, but every citizen of the society would be worse off. Likewise in the reverse direction: it is possible to reduce absolute poverty while increasing relative poverty.

The term *relative poverty* can also be used in a different sense to mean "moderate poverty" – for example, a standard of living or level of income that is high enough to satisfy basic needs (like water, food, clothing, shelter, and basic health care), but still significantly lower than that of the majority of the population under consideration[19] .

Basic needs

See article Basic needs

The basic needs approach is one of the major approaches to the measurement of absolute poverty in developing countries. It attempts to define the absolute minimum resources necessary for long-term physical well-being, usually in terms of consumption goods. The poverty line is then defined as the amount of income required to satisfy those needs. The 'basic needs' approach was introduced by the International Labour Organization's World Employment Conference in 1976.[] [] "Perhaps the high point of the WEP was the World Employment Conference of 1976, which proposed the satisfaction of basic human needs as the overriding objective of national and international development policy. The basic needs approach to development was endorsed by governments and workers' and employers' organizations from all over the world. It influenced the programmes and policies of major multilateral and bilateral development agencies, and was the precursor to the human development approach."[] []

A traditional list of immediate "basic needs" is food (including water), shelter and clothing.[] Many modern lists emphasize the minimum level of consumption of 'basic needs' of not just food, water, and shelter, but also sanitation, education, and health care. Different agencies use different lists.

In 1978 Ghai investigated the literature that criticized the basic needs approach. Critics argued that the basic needs approach lacked scientific rigour; it was consumption-oriented and anti-growth. Some considered it to be "a recipe for perpetuating economic backwardness" and for giving the impression "that poverty elimination is all too easy".[] Amartya Sen focused on 'capabilities' rather than consumption.

In the development discourse, the basic needs model focuses on the measurement of what is believed to be an eradicable level of poverty.

Professor Chris Sarlo, an economist at Nipissing University in North Bay, Canada, and a senior fellow of the Fraser Institute, used Statistics Canada's socio-economic databases, particularly the *Survey of Household Spending* to determine the cost of a list of household necessities. The list includes food, shelter, clothing, health care, personal care, essential furnishings, transportation and communication, laundry, home insurance, and miscellaneous; it assumes that education is provided freely to all residents of Canada. This is calculated for various communities across Canada and adjusted for family size. With this information, he determines the proportion of Canadian households that have insufficient income to afford those necessities. Sarlo introduced his basic needs approach in 1992 in his book entitled *Poverty in Canada* published by the Fraser Institute. Sarlos argued that Statistics Canada's Low Income Cutoff (LICO), which uses a relative poverty line threshold was inadequate. Sarlos' basic needs poverty measure for Canada was updated in 2001 Sarlos' to reflect "new developments in the literature, external critiques of the approach, and better (and more comprehensive) data". For example it included out-of-pocket health-care costs. According to the Sarlo, "the basic-needs approach is partly absolute (the list [of necessities] is limited to items required for long-term physical well-being) and partly relative, reflecting the standards that apply in the individual's own society at the present time." It was updated again in 2006. [] The Fraser Institute's list of necessities for living creditably in Canada includes not only food, shelter, clothing, and health care, but also personal care, furniture, transportation, communication, laundry, and home insurance. It is criticized for not including any entertainment items like cable television, daily newspapers, and tickets to movies or sporting events.[] [] [20] [20] Based on his basic needs poverty threshold, the poverty rate in Canada, the poverty rate has declined from about 12% of Canadian households to about 5% since the 1970s.[21] This is in sharp contrast to Statistic Canada, the Conference Board of Canada, the Organisation for Economic Co-operation and Development (OECD) and UNESCO reports on poverty in Canada. They used the relative poverty measure considered to the most useful for advanced industrial nations like Canada, which Sarlo rejects.

According to a 2008 report by the Organisation for Economic Co-operation and Development (OECD), the rate of poverty in Canada, is among the highest of the OECD member nations, the world's wealthiest industrialized nations.[] There is no official government definition and therefore, measure, for poverty in Canada. However, Raphael Dennis, author of *Poverty in Canada: Implications for Health and Quality of Life*[] [] reported that the

United Nations Development Program (UNDP), the United Nations Children's Fund (UNICEF), the Organisation for Economic Co-operation and Development (OECD) and Canadian poverty researchers[22] [] find that relative poverty is the "most useful measure for ascertaining poverty rates in wealthy developed nations such as Canada".[] [11] [] [12] []

National poverty lines

National estimates are based on population-weighted subgroup estimates from household surveys. Definitions of the poverty line may vary considerably among nations. For example, rich nations generally employ more generous standards of poverty than poor nations. Thus, the numbers are not comparable among countries.

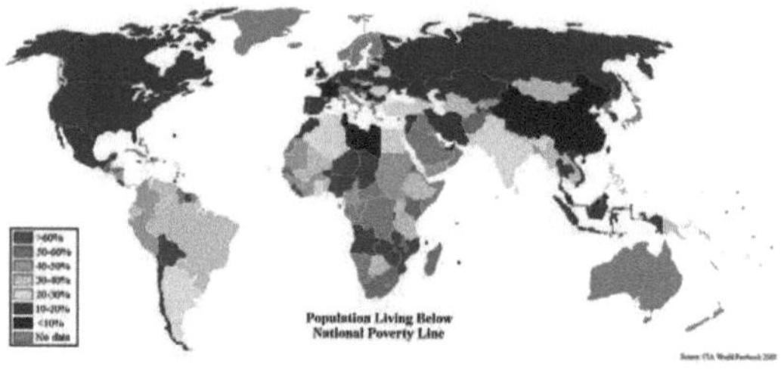

CIA World Factbook-based map showing the percentage of population by country living below that country's official poverty line.

In 2010, in the United States, the poverty threshold for one person under 65 was US$11,344 (annual income); the threshold for a family group of four, including two children, was US$22,133.[] [23] According to the U.S. Census Bureau data released on September 13, 2011, the nation's poverty rate rose to 15.1 percent in 2010.

In the UK, "more than five million people – over a fifth (23 percent) of all employees – were paid less than £6.67 an hour in April 2006. This is based on a low pay rate of 60 percent of full-time median earnings, equivalent to a little over £12,000 a year for a 35-hour working week. In April 2006, a 35-hour week would have earned someone £9,191 a year – before tax or National Insurance".[24] [25]

India's official poverty level, on the other hand, is split according to rural vs. urban thresholds. For urban dwellers, the poverty line is defined as living on less than 538.60 rupees (approximately USD $12) per month, whereas for rural dwellers, it is defined as living on less than 356.35 rupees per month (approximately USD $7.50).[]

Criticisms

Using a poverty threshold is problematic because having an income marginally above it is not substantially different from having an income marginally below it: the negative effects of poverty tend to be continuous rather than discrete, and the same low income affects different people in different ways. To overcome this problem, poverty indices are sometimes used instead; see income inequality metrics.

A poverty threshold relies on a quantitative, or purely numbers-based, measure of income. If other human development-indicators like health and education are used, they must be quantified, which is not a simple (if even achievable) task.

Overstating poverty

In-kind transfers, whether from public or private sources, are not counted when calculating a poverty threshold. For example, if a parent pays the rent on an apartment for an adult son directly to the apartment owner, instead of giving the money to the son to pay the rent, then that money does not count as income to the son. If a church or non-profit organization gives food to an elderly person, the value of the food is not counted as income to the elderly person. Rea Hederman, a senior policy analyst in the Center for Data Analysis at the Heritage Foundation, said

> The official poverty measure counts only monetary income. It considers antipoverty programs such as food stamps, housing assistance, the Earned Income Tax Credit, Medicaid and school lunches, among

> others, "in-kind benefits" – and hence not income. So, despite everything these programs do to relieve poverty, they aren't counted as income when Washington measures the *poverty rate*.[26]

Studies measuring the difference between income before and after taxes and government transfers, however, have found that without these programs poverty would be roughly 30% to 40% higher than the official poverty line indicates, despite many of their benefits not being counted as income.[27] [28]

Further, the U.S. Census Bureau calculates the poverty line the same throughout the U.S. regardless of the cost-of-living in a state or urban area. For instance, the cost-of-living in California, the most populous state, was 1.42 times the U.S. average in 2010 while the cost-of-living in Texas, the second most populous state, was 0.90 (90%) of the U.S. average.

Notes

[1] Ravallion, Martin *Poverty freak: A Guide to Concepts and Methods*. Living Standards Measurement Papers, The World Bank, 1992, p. 25

[2] Hagenaars, Aldi & de Vos, Klaas *The Definition and Measurement of Poverty*. Journal of Human Resources, 1988

[3] Hagenaars, Aldi & van Praag, Bernard *A Synthesis of Poverty Line Definitions*. Review of Income and Wealth, 1985

[4] Sachs, Jeffrey D. *The End of Poverty* 2005, p. 20

[5] Ravallion, Martin; Chen Shaohua & Sangraula, Prem *Dollar a day* The World Bank Economic Review, 23, 2, 2009, pp. 163-184

[6] {Poverty Lines-Martin Ravallion, in The New Palgrave Dictionary of Economics, 2nd Edition, London: Palgrave Macmillan}

[7] Alan Gillie, 'The Origin of the Poverty Line', Economic History Review, XLIX/4 (1996), 726

[8] David Boyle - The Tyranny of Numbers p.116

[9] page 298

[11] Child poverty in rich nations: Report card no. 6 (Report). Innocenti Research Centre. 2005.

[12] Human development report: Capacity development: Empowering people and institutions (Report). Geneva: United Nations Development Program. 2008.

[13] Jonathan Bradshaw; Yekaterina Chzhen; Gill Main; Bruno Martorano; Leonardo Menchini; Chris de Neubourg (January 2012) (PDF). [www.unicef-irc.org/publications/pdf/iwp_2012_01.pdf Relative Income Poverty among Children in Rich Countries] (Report). Innocenti Working Paper. Florence, Italy: UNICEF Innocenti Research Centre. ISSN 1014-7837. www.unicef-irc.org/publications/pdf/iwp_2012_01.pdf.

[14] A League Table of Child Poverty in Rich Nations (Report). Innocenti Report Card No.1. Florence, Italy: UNICEF Innocenti Research Centre.

[15] Peter Adamson (2012) (PDF). Measuring child poverty: New league tables of child poverty in the world's rich countries (http://www.unicef.ca/sites/default/files/imce_uploads/DISCOVER/OUR WORK/ADVOCACY/DOMESTIC/POLICY ADVOCACY/DOCS/unicefreportcard10-eng.pdf) (Report). UNICEF Innocenti Research Centre Report Card. Florence, Italy. .

[16] Minority [Republican] views, p. 46 in U.S. Congress, Report of the Joint Economic Committee on the January 1964 Economic Report of the President with Minority and Additional Views (Report). Washington, D.C.: US Government Printing Office. January 1964.

[17] Friedman, Rose. D. (1965). *"Poverty: Definition and Perspective"*. *American Enterprise Institute for Public Policy Research* (Washington, D.C.).

[19] http://www.caledonia.org.uk/papers/Who-and-Where-are-the-Poor.pdf

[20] Chris Sarlo (23 July 2001). Measuring Poverty in Canada (Report). Fraser Institute.

[21] Poverty in Canada: 2006 Update (http://www.fraserinstitute.org/COMMERCE.WEB/product_files/PovertyinCanada2006.pdf)

[22] The Conference Board of Canada "uses the OECD's relative measure of child poverty, which calculates the proportion of children living in households where disposable income is less than 50 per cent of the median in each country." The Conference Board 2013 cautioned that Canada's high poverty rate, ranks among the worst of the 17 countries they compared. "Canada's child poverty rate was 15.1 per cent, up from 12.8 per cent in the mid-1990s. Only the United States ranked lower.

[24] Working out of Poverty: A study of the low paid and the working poor by Graeme Cooke and Kayte Lawton

[25] IPPR Article: "Government must rescue 'forgotten million children' in poverty" (http://www.ippr.org.uk/pressreleases/?id=2965)

[26] "Poor poverty yardsticks" (http://www.heritage.org/Press/Commentary/ed090806b.cfm), Rea S. Hederman, Jr., September 8, 2006, The Heritage Foundation

[27] Kenworthy, L. (1999). Do social-welfare policies reduce poverty? A cross-national assessment. *Social Forces, 77*(3), 1119-1139. doi:10.1093/sf/77.3.1119

[28] Bradley, D., Huber, E., Moller, S., Nielson, F. & Stephens, J. D. (2003). Determinants of relative poverty in advanced capitalist democracies. (http://www.jstor.org/stable/3088901) *American Sociological Review, 68*(3), 22-51.

See also

- Asset poverty
- Income deficit
- List of countries by percentage of population living in poverty
- Living wage
- Minimum wage
- Measuring poverty
- UN Millennium Development Goals

References

Further reading

- Alan Gillie, "The Origin of the Poverty Line", *Economic History Review,* XLIX/4 (1996), 726

External links

- History of the U.S. Poverty Line (http://eesc.orst.edu/agcomwebfile/edmat/html/em/em8743/part1/officialgovt.html) by Tom Gentle, Oregon State University.
- United States Department of Health and Human Services Poverty Guidelines, Research, and Measurement (http://aspe.hhs.gov/poverty/index.shtml)
- 2007 United States Department of Health and Human Services Poverty Guidelines (http://aspe.hhs.gov/poverty/07poverty.shtml)
- Debraj Ray 1998, *Development Economics*, Princeton University Press, ISBN 0-691-01706-9.
- World Summit for Social Development Agreements (http://www.un.org/esa/socdev/wssd/agreements/index.html), United Nations
- The "elimination" of poverty (http://www.inclusivedemocracy.org/journal/vol4/vol4_no1_takis_poverty.htm), Takis Fotopoulos, *The International Journal of Inclusive Democracy* (http://www.inclusivedemocracy.org/journal/), vol.4, no.1 (January 2008).

Time_zone

A **time zone** is a region on Earth that has a uniform standard time for legal, commercial, and social purposes. It is convenient for areas in close commercial or other communication to keep the same time, so time zones tend to follow the boundaries of countries and their subdivisions.

Most of the time zones on land are offset from Coordinated Universal Time (UTC) by a whole number of hours (UTC−12 to UTC+14), but a few are offset by 30 or 45 minutes. Some higher latitude countries use daylight saving time for part of the year, typically by changing clocks by an hour. Many land time zones are skewed toward the west of the corresponding nautical time zones. This also creates a permanent daylight saving time effect.

History

Early timekeeping

Before the invention of clocks, people marked the time of day with apparent solar time (also called "true" solar time) – for example, the time on a sundial – which was typically different for every settlement.

When well-regulated mechanical clocks became widespread in the early 19th century, each city began to use some local mean solar time. Apparent and mean solar time can differ by up to around 15 minutes (as described by the equation of time) due to the non-circular shape of the Earth's orbit around the sun and the tilt of the Earth's axis. Mean solar time has days of equal length, and the difference between the two averages to zero after a year.

Greenwich Mean Time (GMT) was established in 1675 when the Royal Observatory was built as an aid to mariners to determine longitude at sea, providing a standard reference time when each city in England kept a different local time.

Railroad time

Local solar time became increasingly awkward as railways and telecommunications improved, because clocks differed between places by an amount corresponding to the difference in their geographical longitude, which varied by four minutes for every degree of longitude. The difference between New York and Boston is about two degrees or 8 minutes, the difference between Sydney and Melbourne, Australia, is about 7 degrees or 28 minutes. Bristol is 2°35′ W(est) of Greenwich (East London), so when it is noon in Bristol, it is about 10 minutes past noon in London.[1] The use of time zones smooths out these differences.

Plaque commemorating the Railway General Time Convention of 1883 in North America

The first adaptation of a standard time in the world was established on December 1, 1847, in Great Britain by railway companies using GMT kept by portable chronometers. The first of these companies to adopt standard time was the Great Western Railway (GWR) in November 1840. This quickly became known as Railway Time. About August 23, 1852, time signals were first transmitted by telegraph from the Royal Observatory, Greenwich. Even though 98% of Great Britain's public clocks were using GMT by 1855, it was not made Britain's legal time until August 2, 1880. Some old British clocks from this period have two minute hands—one for the local time, one for GMT.[2]

The increase in worldwide communication had further increased the need for interacting parties to communicate mutually comprehensible time references to one another. The problem of differing local times could be solved across larger areas by synchronizing clocks worldwide, but in many places the local time would then differ markedly from the solar time to which people were accustomed. Time zones were a compromise, relaxing the complex geographic dependence while still allowing local time to approximate the mean solar time.

On November 2, 1868, the then-British colony of New Zealand officially adopted a standard time to be observed throughout the colony, and was perhaps the first country to do so. It was based on the longitude 172°30′ East of Greenwich, that is 11 hours 30 minutes ahead of GMT. This standard was known as New Zealand Mean Time.

Timekeeping on the American railroads in the mid-19th century was somewhat confused. Each railroad used its own standard time, usually based on the local time of its headquarters or most important terminus, and the railroad's train schedules were published using its own time. Some major railroad junctions served by several different railroads had a separate clock for each railroad, each showing a different time; the main station in Pittsburgh, Pennsylvania, for example, kept six different times.

Charles F. Dowd proposed a system of one-hour standard time zones for American railroads about 1863, although he published nothing on the matter at that time and did not consult railroad officials until 1869. In 1870, he proposed four ideal time zones (having north–south borders), the first centered on Washington, D.C., but by 1872 the first was centered 75°W of Greenwich, with geographic borders (for example, sections of the Appalachian Mountains). Dowd's system was never accepted by American railroads. Instead, U.S. and Canadian railroads implemented a version proposed by William F. Allen, the editor of the *Traveler's Official Railway Guide*.[3] The borders of its time zones ran through railroad stations, often in major cities. For example, the border between its Eastern and Central time zones ran through Detroit, Buffalo, Pittsburgh, Atlanta, and Charleston. It was inaugurated on Sunday, November 18, 1883, also called "The Day of Two Noons",[4] when each railroad station clock was reset as standard-time noon was reached within each time zone. The zones were named Intercolonial, Eastern, Central, Mountain, and Pacific. Within one year, 85% of all cities with populations over 10,000, about 200 cities, were using standard time.[5] A notable exception was Detroit (which is about half-way between the meridians of eastern time and central time), which kept local time until 1900, then tried Central Standard Time, local mean time, and Eastern Standard Time before a May 1915 ordinance settled on EST and was ratified by popular vote in August 1916. The confusion of times came to an end when Standard zone time was formally adopted by the U.S. Congress in the Standard Time Act of March 19, 1918.

Worldwide time zones

Although the first person to propose a worldwide system of time zones was Italian mathematician Quirico Filopanti in his book *Miranda!* published in 1858, his idea was unknown outside the pages of his book until long after his death, so it did not influence the adoption of time zones during the 19th century. He proposed 24 hourly time zones, which he called "longitudinal days", the first centered on the meridian of Rome. He also proposed a universal time to be used in astronomy and telegraphy.[6] [7]

Scottish-born Canadian Sir Sandford Fleming proposed a worldwide system of time zones in 1879. He advocated his system at several international conferences, thus is widely credited with their invention. In 1876, his first proposal was for a global 24-hour clock, conceptually located at the center of the Earth and not linked to any surface meridian. In 1879 he specified that his universal day would begin at the anti-meridian of Greenwich (180th meridian), while conceding that hourly time zones might have some limited local use. He also proposed his system at the International Meridian Conference in October 1884, but it did not adopt his time zones because they were not within its purview. The conference did adopt a universal day of 24 hours beginning at Greenwich midnight, but specified that it "shall not interfere with the use of local or standard time where desirable".

By about 1900, almost all time on Earth was in the form of standard time zones, only some of which used an hourly offset from GMT. Many applied the time at a local astronomical observatory to an entire country, without any reference to GMT. It took many decades before all time on Earth was in the form of time zones referred to some "standard offset" from GMT/UTC. Most major countries had adopted hourly time zones by 1929. Nepal was the last country to adopt a standard offset, shifting slightly to UTC+5:45 in 1986.

Today, all nations use standard time zones for secular purposes, but they do not all apply the concept as originally conceived. Newfoundland, India, Iran, Afghanistan, Venezuela, Burma, the Marquesas, as well as parts of Australia use half-hour deviations from standard time, and some nations, such as Nepal, and some provinces, such as the Chatham Islands, use quarter-hour deviations. Some countries, most notably China and India, use a single time zone, even though the extent of their territory far exceeds 15° of longitude. Before 1949, China used five time zones (see Time in China).

Definition

Before 1972, all time zones were specified as an offset from Greenwich Mean Time (GMT), which was the mean solar time at the meridian passing through the Royal Observatory in Greenwich, London. Since 1972 all official time services have broadcast radio time signals synchronized to UTC, a form of atomic time that includes leap seconds to keep it within 0.9 seconds of this former GMT, now called UT1. Many countries now legally define their standard time relative to UTC, although some still legally refer to GMT, including the United Kingdom itself. UTC, also called Zulu time, is used everywhere on Earth by astronomers and others who need to state the time of an event unambiguously.

Time zones are based on Greenwich Mean Time (GMT), the mean solar time at longitude 0° (the Prime Meridian). The definition of GMT was recently changed – it was previously the same as UT1, a mean solar time calculated directly from the rotation of the Earth. As the rate of rotation of the Earth is not constant, the time derived from atomic clocks was adjusted to closely match UT1. In January 1972, however, the length of the second in both Greenwich Mean Time and atomic time was equalized. The readings of participating atomic clocks are averaged out to give a uniform time scale.

Because the length of the average day is a small fraction of a second more than 24 hours (slightly more than 86400 seconds), leap seconds are periodically inserted into Greenwich Mean Time to make it approximate to UT1. This new time system is also called Coordinated Universal Time (UTC). Leap seconds are inserted to keep UTC within 0.9 seconds of UT1. Because the Earth's rotation is gradually slowing, leap seconds will need to be added more frequently in the future. However, from one year to the next the rotation rate is slightly irregular, so leap seconds are not added unless observations of Earth's rotation show that one is required. In this way, local times will continue to stay close to mean solar time and the effects of variations in Earth's rotation rate will be confined to simple step changes relative to the uniform time scale (International Atomic Time or TAI). All local times differ from TAI by an integral number of seconds. With the implementation of UTC, nations began to use it in the definition of their time zones. As of 2005, most nations had altered the definition of local time in this way.

In the UK, this involved redefining Greenwich Mean Time to make it the same as UTC.[8] British Summer Time (BST) is still one hour in advance of Greenwich Mean Time and is therefore also one hour in advance of Coordinated Universal Time. Thus Greenwich Mean Time is the local time at the Royal Observatory, Greenwich between 0100 hours GMT on the last Sunday in October and 0100 hours GMT on the last Sunday in March. Similar circumstances apply in many other places.

Leap seconds are considered by many to be a nuisance, and ways to abolish them are being considered. This means letting the time difference accumulate. One suggestion is to insert a "leap-hour" in about 5,000 years. For more on this discussion read Proposal to abolish leap seconds.

Notation of time

ISO 8601

UTC

If the time is in Coordinated Universal Time (UTC), add a "Z" directly after the time without a space. "Z" is the zone designator for the zero UTC offset. "09:30 UTC" is therefore represented as "09:30Z" or "0930Z". "14:45:15 UTC" would be "14:45:15Z" or "144515Z".

UTC time is also known as "Zulu" time, since "Zulu" is the ICAO spelling alphabet code word for "Z".

Offsets from UTC

Offsets from UTC are written in the format ±[hh]:[mm], ±[hh][mm], or ±[hh]. So if the time being described is one hour ahead of UTC (such as the time in Berlin during the winter), the zone designator would be "+01:00", "+0100", or simply "+01". This is appended to the time in the same way that 'Z' was above. The offset from UTC changes with daylight saving time, e.g. a time offset in Chicago, which is in the North American Central Time Zone, would be "−06:00" for the winter (Central Standard Time) and "−05:00" for the summer.

Abbreviations

Time zones are often represented by abbreviations such as "EST, WST, CST" but these are not part of the international time and date standard ISO 8601 and their use as sole designator for a time zone is not recommended. Such designations can be ambiguous. For example, "BST", which is British Summer Time, was renamed "British Standard Time" between 1968 and 1971 when Central European Time was in force because legislators objected to calling it Central European Time. The same legislation affirmed that the Standard Time within the United Kingdom was, and would continue to be, Greenwich Mean Time.

UTC offsets worldwide

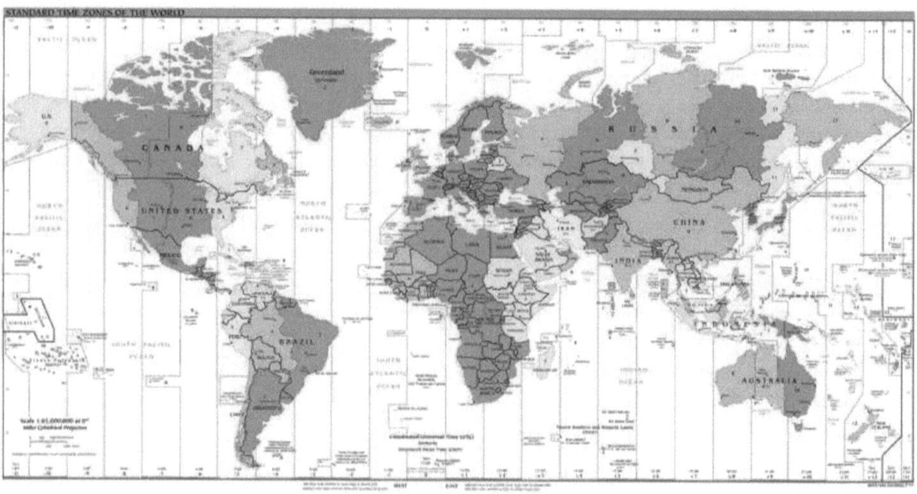

Time zones map (Last update: March 2013)

List of UTC offsets

These examples give the local time at various locations around the world when daylight saving time is not in effect:

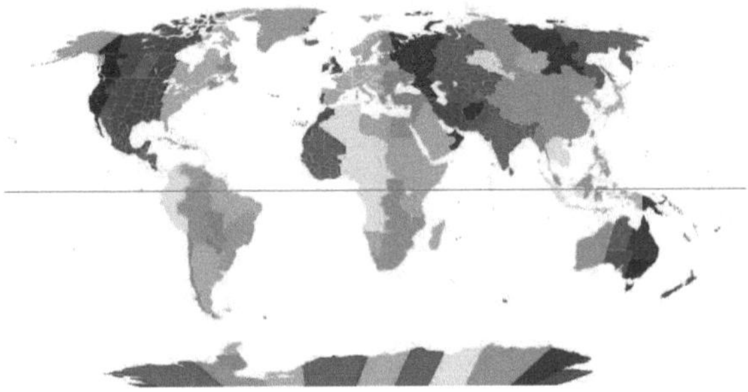

XX = ISO 3166-1 alpha-2 country code, XX- = parts of the country, N = North, S = South, UTC = Universal Coordinated Time, DST = Daylight Saving Time

Time zone	Example time (ISO 8601 notation)	Example locations	Example locations that in summer use DST
UTC−12:00	2013-08-12T03:56:03-12:00	Baker Island, Howland Island (both uninhabited)	
UTC−11:00	2013-08-12T04:56:03-11:00	American Samoa, Niue	
UTC−10:00	2013-08-12T05:56:03-10:00	United States (Hawaii)	Most of French Polynesia, United States (Aleutian Islands)
UTC−09:30	2013-08-12T06:26:03-09:30	Marquesas Islands	
UTC−09:00	2013-08-12T06:56:03-09:00	Gambier Islands	United States (most of Alaska)
UTC−08:00	2013-08-12T07:56:03-08:00		Canada (most of British Columbia), Mexico (Baja California), United States (California, most of Nevada, most of Oregon, Washington (state))
UTC−07:00	2013-08-12T08:56:03-07:00	Canada (northeastern British Columbia), Mexico (Sonora), United States (Arizona)	Canada (Alberta), Mexico (Chihuahua), United States (Colorado)
UTC−06:00	2013-08-12T09:56:03-06:00	Canada (almost all of Saskatchewan), Costa Rica, El Salvador, Ecuador (Galápagos Islands), Guatemala, Honduras, Mexico (most), Nicaragua,	Canada (Manitoba), United States (Illinois, most of Texas)
UTC−05:00	2013-08-12T10:56:03-05:00	Colombia, Cuba, Ecuador (continental), Jamaica, Panama, Peru	Canada (most of Ontario, most of Quebec), Haiti, United States (most of Florida, Georgia, Massachusetts, most of Michigan, most of Indiana, New York, North Carolina, Ohio, Washington, D.C.)
UTC−04:30	2013-08-12T11:26:03-04:30	Venezuela	
UTC−04:00	2013-08-12T11:56:03-04:00	Bolivia, Brazil (Amazonas), Chile (continental), Dominican Republic, Canada (Nova Scotia), Puerto Rico, Trinidad and Tobago	Falkland Islands
UTC−03:30	2013-08-12T12:26:03-03:30		Canada (island of Newfoundland and southern Labrador)
UTC−03:00	2013-08-12T12:56:03-03:00	Argentina, Paraguay	Brazil (Federal District, Rio de Janeiro, São Paulo), most of Greenland, Uruguay
UTC−02:00	2013-08-12T13:56:03-02:00	Brazil (Fernando de Noronha), South Georgia and the South Sandwich Islands	

UTC−01:00	2013-08-12T14:56:03-01:00	Portugal (Azores), Cape Verde	
UTC±00:00	2013-08-12T15:56:03+00:00	Côte d'Ivoire, Faroe Islands, Ghana, Iceland, Senegal	Ireland, Portugal (continental and Madeira), Spain (Canary Islands), Morocco, United Kingdom, Western Sahara
UTC+01:00	2013-08-12T16:56:03+01:00	Algeria, Angola, Benin, Cameroon, Gabon, Niger, Nigeria, Democratic Republic of the Congo (west), Tunisia	Albania, Austria, Belgium, Bosnia and Herzegovina, Spain (continental), Croatia, Czech Republic, Denmark, Germany, Hungary, Italy, Libya, Malta, Kosovo, Macedonia, France (metropolitan), Namibia, the Netherlands, Norway, Poland, Serbia, Slovakia, Slovenia, Sweden, Switzerland
UTC+02:00	2013-08-12T17:56:03+02:00	Egypt, Malawi, Mozambique, South Africa, Swaziland, Zambia, Zimbabwe	Bulgaria, Cyprus, Estonia, Finland, Greece, Israel, Latvia, Lebanon, Lithuania, Moldova, Northern Cyprus, Palestine, Romania, Syria, Transnistria, Turkey, Ukraine
UTC+03:00	2013-08-12T18:56:03+03:00	Belarus, Djibouti, Eritrea, Ethiopia, Iraq, Jordan, Kenya, Madagascar, Russia (Kaliningrad Oblast), Saudi Arabia, South Sudan, Sudan, Somalia, Somaliland, Tanzania, Uganda, Yemen	
UTC+03:30	2013-08-12T19:26:03+03:30		Iran
UTC+04:00	2013-08-12T19:56:03+04:00	Abkhazia, Armenia, Azerbaijan, Georgia, Mauritius, Oman, Russia (European), Seychelles, South Ossetia, United Arab Emirates	
UTC+04:30	2013-08-12T20:26:03+04:30	Afghanistan	
UTC+05:00	2013-08-12T20:56:03+05:00	Kazakhstan (west), Maldives, Pakistan, Uzbekistan	
UTC+05:30	2013-08-12T21:26:03+05:30	India, Sri Lanka	
UTC+05:45	2013-08-12T21:41:03+05:45	Nepal	
UTC+06:00	2013-08-12T21:56:03+06:00	Kazakhstan (most), Bangladesh, Bhutan, Russia (Ural: Sverdlovsk Oblast, Chelyabinsk Oblast)	
UTC+06:30	2013-08-12T22:26:03+06:30	Cocos Islands, Myanmar	
UTC+07:00	2013-08-12T22:56:03+07:00	Jakarta, Russia (Novosibirsk Oblast), Thailand, Vietnam	
UTC+08:00	2013-08-12T23:56:03+08:00	China, Hong Kong, Russia (Krasnoyarsk Krai), Malaysia, Philippines, Singapore, Taiwan, most of Mongolia, Western Australia	
UTC+08:45	2013-08-13T00:41:03+08:45	Australia (Eucla)	
UTC+09:00	2013-08-13T00:56:03+09:00	East Timor, Russia (Irkutsk Oblast), Japan, North Korea, South Korea	
UTC+09:30	2013-08-13T01:26:03+09:30	Australia (Northern Territory)	Australia (South Australia)
UTC+10:00	2013-08-13T01:56:03+10:00	Russia (Zabaykalsky Krai), Papua New Guinea, Australia (Queensland)	Australia (New South Wales, Tasmania, Victoria)
UTC+10:30	2013-08-13T02:26:03+10:30		Lord Howe Island
UTC+11:00	2013-08-13T02:56:03+11:00	New Caledonia, Russia (Primorsky Krai), Solomon Islands	
UTC+11:30	2013-08-13T03:26:03+11:30	Norfolk Island	

UTC+12:00	2013-08-13T03:56:03+12:00	Kiribati (Gilbert Islands), Fiji, Russia (Kamchatka Krai)	New Zealand
UTC+12:45	2013-08-13T04:41:03+12:45		Chatham Islands
UTC+13:00	2013-08-13T04:56:03+13:00	Kiribati (Phoenix Islands), Tonga, Tokelau	Samoa
UTC+14:00	2013-08-13T05:56:03+14:00	Kiribati (Line Islands)	

Where the adjustment for time zones results in a time at the other side of midnight from UTC, then the date at the location is one day later or earlier.

Some examples when UTC is 23:00 on Monday when or where daylight saving time is not in effect:

- Cairo, Egypt: UTC+02; 01:00 on Tuesday
- Wellington, New Zealand: UTC+12; 11:00 on Tuesday

Some examples when UTC is 02:00 on Tuesday when or where daylight saving time is not in effect:

- Honolulu, Hawaii, United States: UTC−10; 16:00 on Monday
- Toronto, Ontario, Canada: UTC−05; 21:00 on Monday

The time-zone adjustment for a specific location may vary because of daylight saving time. For example New Zealand, which is usually UTC+12, observes a one-hour daylight saving time adjustment during the Southern Hemisphere summer, resulting in a local time of UTC+13.

Time zone conversions

Conversion between time zones obeys the relationship

"time in zone A" − "UTC offset for zone A" = "time in zone B" − "UTC offset for zone B",

in which each side of the equation is equivalent to UTC. (The more familiar term "UTC offset" is used here rather than the term "zone designator" used by the standard.)

The conversion equation can be rearranged to

"time in zone B" = "time in zone A" − "UTC offset for zone A" + "UTC offset for zone B".

For example, what time is it in Los Angeles (UTC offset= −08) when the New York Stock Exchange opens at 09:30 (−05)?

time in Los Angeles = 09:30 − (−05:00) + (−08:00) = 06:30.

In Delhi (UTC offset= +5:30), the New York Stock Exchange opens at

time in Delhi = 09:30 − (−05:00) + (+5:30) = 20:00.

These calculations become more complicated near a daylight saving boundary (because the UTC offset for zone X is a function of the UTC time).

Nautical time zones

Since the 1920s a nautical standard time system has been in operation for ships on the high seas. Nautical time zones are an ideal form of the terrestrial time zone system. Under the system, a time change of one hour is required for each change of longitude by 15°. The 15° gore that is offset from GMT or UT1 (not UTC) by twelve hours is bisected by the nautical date line into two 7.5° gores that differ from GMT by ±12 hours. A nautical date line is implied but not explicitly drawn on time zone maps. It follows the 180th meridian except where it is interrupted by territorial waters adjacent to land, forming gaps: it is a pole-to-pole dashed line.[9] [10] [11]

A ship within the territorial waters of any nation would use that nation's standard time, but would revert to nautical standard time upon leaving its territorial waters. The captain is permitted to change the ship's clocks at a time of the captain's choice following the ship's entry into another time zone. The captain often chooses midnight. Ships going in shuttle traffic over a time zone border often keeps the same time zone all the time, to avoid confusion about work, meal and shop opening hours. Still the time table for port calls must follow the land time zone.

Skewing of zones

Ideal time zones, such as nautical time zones, are based on the mean solar time of a particular meridian located in the middle of that zone with boundaries located 7.5 degrees east and west of the meridian. In practice, zone boundaries are often drawn much farther to the west with often irregular boundaries, and some locations base their time on meridians located far to the east.

Difference between sun time and clock time during daylight saving time:

For example, even though the Prime Meridian (0°) passes through Spain and France, they use the mean solar time of 15 degrees east (Central European Time) rather than 0 degrees (Greenwich Mean Time). France previously used GMT, but was switched to CET (Central European Time) during the German occupation of the country during World War II and did not switch back after the war.[12] Similarly, prior to World War II, the Netherlands observed "Amsterdam Time", which was twenty minutes ahead of Greenwich Mean Time. They were obliged to follow German time during the war, and kept it thereafter. In the mid 1970s the Netherlands, as with other European states, began observing daylight saving (summer) time.

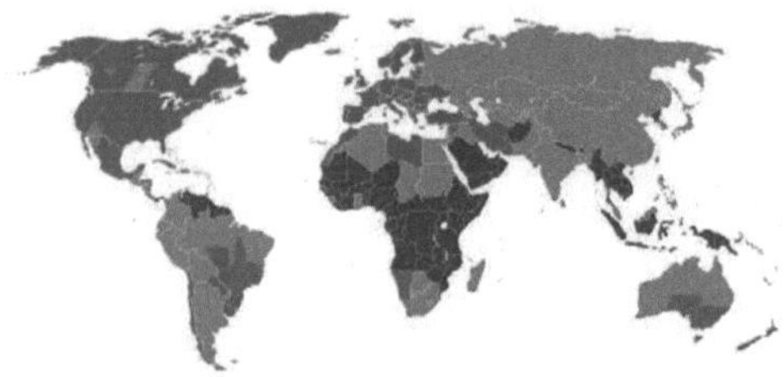

DST observed DST no longer observed DST never observed

There is a tendency to draw time zone boundaries far to the west of their meridians. Many of these locations also use daylight saving time. As a result, in the summer, solar noon in the Spanish town of Muxia occurs at 2:37 pm by the clock. This area of Spain never experiences sunset before 6:00 pm local time even in midwinter, despite its lying more than 40 degrees north of the equator. Near the summer solstice, Muxia has sunset times similar to those of Stockholm, which is in the same time zone and 16 degrees further north.

A more extreme example is Nome, Alaska, which is at 165°24′W longitude—just west of center of the idealized Samoa Time Zone (165°W). Nevertheless, Nome observes Alaska Time (135°W) with DST so it is slightly more than two hours ahead of the sun in winter and over three in summer.[13] Kotzebue, Alaska, also near the same meridian but north of the Arctic Circle, has an annual event on 9 August to celebrate *two* sunsets in the same 24-hour day, one shortly after midnight at the start of the day, and the other shortly before midnight at the end of the day.

Also, China extends as far west as 73°34′E, but all parts of it use UTC+08:00 (120°E), so solar "noon" can occur as late as 15:00 in western portions of China such as Xinjiang and Tibet.

Daylight saving time

Many countries, and sometimes just certain regions of countries, adopt daylight saving time (also known as "Summer Time") during part of the year. This typically involves advancing clocks by an hour near the start of spring and adjusting back in autumn ("spring" forward, "fall" back). Modern DST was first proposed in 1907 and was in widespread use in 1916 as a wartime measure aimed at conserving coal. Despite controversy, many countries have used it off and on since then; details vary by location and change occasionally. Most countries around the equator do not observe daylight saving time, since the seasonal difference in sunlight is minimal.

Additional information

- France has twelve time zones including those of Metropolitan France, French Guiana and numerous islands, inhabited and uninhabited. Russia has nine time zones (and used to have 11 time zones before March 2010), eight contiguous zones plus Kaliningrad exclave on the Baltic Sea. The United States has ten time zones (nine official plus that for Wake Island and its Antarctic stations; no official time zone is specified for uninhabited Howland Island and Baker Island). Australia has nine time zones (one unofficial and three official on the mainland plus four for its territories and one more for an Antarctic station not included in other time zones). The United Kingdom has eight time zones for itself and its overseas territories. Canada has six official time zones. The Danish Realm has five time zones.
- In terms of area, China is the largest country with only one time zone (UTC+08). China also has the widest spanning time zone. Before 1949, China was separated into five time zones.
- Stations in Antarctica generally keep the time of their supply bases, thus both the Amundsen-Scott South Pole Station (U.S.) and McMurdo Station (U.S.) use New Zealand time (UTC+12 southern winter, UTC+13 southern summer).
- The 27°N latitude passes back and forth across time zones in South Asia. Pakistan: +05:00, India +05:30, Nepal +05:45, +05:30, Myanmar +06:30. This switching was more odd in 2002, when Pakistan enabled daylight saving time. Thus from west to east, time zones were: +06:00, +05:30, +05:45, +05:30, +08:00, +06:00, +05:30 and +06:30.
- Because the earliest and latest time zones are 26 hours apart, any given calendar date exists at some point on the globe for 50 hours. For example, April 11 begins in time zone UTC+14 at 10:00 UTC April 10, and ends in time zone UTC−12 at 12:00 UTC April 12.
- There are 22 places where three or more time zones meet, for instance at the tri-country border of Finland, Norway and Russia. 28 countries present such triple points, with China being the most present (in 13 of the 22 triple points). Then come India (7), Russia, India and Afghanistan (4).
- There are 40 time zones. This is due to fractional hour offsets and zones with offsets larger than 12 hours near the International Date Line as well as one unofficial zone in Australia. See the list of time zones.
- The largest time gap along a political border is the 3.5 hour gap along the border of China (UTC+08) and Afghanistan (UTC+04:30).
- One of the most unusual time zones is the Australian Central Western Time zone (CWST), which is a small strip of Western Australia from the border of South Australia west to 125.5°E, just before Caiguna. It is 8¾ hours ahead of UTC (UTC+08:45) and covers an area of about 35,000 km^2, larger than Belgium, but has a population of about 200. Although unofficial, it is universally respected in the area—without it, the time gap in standard time at 129°E (the WA/SA border) would be 1.5 hours. See Time in Australia.

Computer systems and the Internet

Computer operating systems include the necessary support for working with all (or almost all) possible local times based on the various time zones. Internally, operating systems typically use UTC as their basic time-keeping standard, while providing services for converting local times to and from UTC, and also the ability to automatically change local time conversions at the start and end of daylight saving time in the various time zones. (See the article on daylight saving time for more details on this aspect).

Web servers presenting web pages primarily for an audience in a single time zone or a limited range of time zones typically show times as a local time, perhaps with UTC time in brackets. More internationally-oriented websites may show times in UTC only or using an arbitrary time zone. For example the international English-language version of CNN includes GMT and Hong Kong Time,[14] whilst the US version shows Eastern Time.[15] US Eastern Time and Pacific Time are also used fairly commonly on many US-based English-language websites with global readership. The format is typically based in the W3C Note "datetime".

Email systems and other messaging systems (IRC chat, etc.)[16] time-stamp messages using UTC, or else include the sender's time zone as part of the message, allowing the receiving program to display the message's date and time of sending in the recipient's local time.

Database records that include a time stamp typically use UTC, especially when the database is part of a system that spans multiple time zones. The use of local time for time-stamping records is not recommended for time zones that implement daylight saving time due to the fact that once a year there is a one hour period when local times are ambiguous.

Operating systems

Unix

Most Unix-like systems, including Linux and Mac OS X, keep system time as UTC (Coordinated Universal Time). Rather than having a single time zone set for the whole computer, timezone offsets can vary for different processes. Standard library routines are used to calculate the local time based on the current timezone, normally supplied to processes through the TZ environment variable. This allows users in multiple timezones to use the same computer, with their respective local times displayed correctly to each user. Time zone information most commonly comes from the IANA time zone database. In fact, many systems, including anything using the GNU C Library, can make use of this database.

Microsoft Windows

Windows-based computer systems prior to Windows 2000 used local time, but Windows 2000 and later can use UTC as the basic system time.[17] The system registry contains time zone information that includes the offset from UTC and rules that indicate the start and end dates for daylight saving in each zone. Interaction with the user normally uses local time, and application software is able to calculate the time in various zones. Terminal Servers allow remote computers to redirect their time zone settings to the Terminal Server so that users see the correct time for their time zone in their desktop/application sessions. Terminal Services uses the server base time on the Terminal Server and the client time zone information to calculate the time in the session.

Programming languages

Java

While most application software will use the underlying operating system for timezone information, the Java Platform, from version 1.3.1, has maintained its own timezone database. This database will need to be updated whenever timezone rules change. Sun provides an updater tool for this purpose.[18]

As an alternative to the timezone information bundled with the Java Platform, programmers may choose to use the Joda-Time library.[19] This library includes its own timezone data based on the IANA time zone database.[20]

JavaScript

There is very little in the way of timezone support for JavaScript. Essentially the programmer has to extract the UTC offset by instantiating a time object, getting a GMT time from it, and differencing the two. This does not provide a solution for daylight saving variations.

PHP

The DateTime objects and related functions have been compiled into the PHP core since 5.2. This includes the ability to get and set the default script timezone, and DateTime is aware of its own timezone internally. PHP.net provides extensive documentation on this.[21] As noted there, the most current timezone database can be implemented via the PECL timezonedb.

Python

The standard module datetime stores and operates on the timezone information class tzinfo. The third party pytz module provides access to the full IANA time zone database.[22] Negated time zone offset in seconds is stored time.timezone and time.altzone attributes.

Smalltalk

Each Smalltalk dialect comes with its own built-in classes for dates, times and timestamps, only a few of which implement the DateAndTime and Duration classes as specified by the ANSI Smalltalk Standard. VisualWorks provides a TimeZone class that supports up to two annually recurring offset transitions, which are assumed to apply to all years (same behavior as Windows time zones). Squeak provides a Timezone class that does not support any offset transitions. Dolphin Smalltalk does not support time zones at all.

For full support of the tz database (zoneinfo) in a Smalltalk application (including support for any number of annually recurring offset transitions, and support for different intra-year offset transition rules in different years) the third-party, open-source, ANSI-Smalltalk-compliant Chronos Date/Time Library is available for use with any of the following Smalltalk dialects: VisualWorks, Squeak, Gemstone, or Dolphin.[23]

Databases

Some databases allow storage of a datetime type having time zone information. The SQL standard defines two standard time data types:

TIMESTAMP WITH TIME ZONE

TIMESTAMP WITHOUT TIME ZONE

However, the standard has a somewhat naive understanding of time zones. It generally assumes a time zone can be specified by a simple offset from GMT. This causes problems when trying to do arithmetic on dates which span daylight saving time transitions or which span political changes in time zone rules.

Oracle

Oracle Database is configured with a database time zone, and connecting clients are configured with session time zones. Oracle Database uses two data types to store time zone information:

TIMESTAMP WITH TIME ZONE

Stores date and time information with the offset from UTC

TIMESTAMP WITH LOCAL TIME ZONE

Stores date and time information with respect to the dbtimezone (which cannot be changed so long as there is a column in the db of this type), automatically adjusting the date and time from the stored time zone to the client's session time zone.

PostgreSQL

PostgreSQL uses the standard SQL data types but tries to impose an interpretation which avoids the problems described above.

TIMESTAMP WITH TIME ZONE

Stores date and time in UTC and converts to the client's local time zone (which could be different for each client) for display purposes and conversion to other types.

TIMESTAMP WITHOUT TIME ZONE

Stores date and time without any conversion on input or output. When converting to a TIMESTAMP WITH TIME ZONE, interprets it according to the client's local time zone.

TIME WITH TIME ZONE

Stores time of day together with a UTC offset in which it is to be interpreted.

TIME WITHOUT TIME ZONE

Stores time of day without any time zone specification.

Time zones in outer space

Orbiting spacecraft typically experience many sunrises and sunsets in a 24-hour period, or in the case of Apollo program astronauts travelling to the moon, none. Thus it is not possible to calibrate time zones with respect to the sun, and still respect a 24-hour sleep/wake cycle. A common practice for space exploration is to use the Earth-based time zone of the launch site or mission control. This keeps the sleeping cycles of the crew and controllers in sync. The International Space Station normally uses Coordinated Universal Time (UTC).

Timekeeping on Mars can be more complex, since the planet has a solar day of approximately 24 hours and 39 minutes, known as a sol. Earth controllers for some Mars missions have synchronized their sleep/wake cycles with the Martian day, because solar-powered rover activity on the surface was tied to periods of light and dark. The difference in day length caused the sleep/wake cycles to slowly drift with respect to the day/night cycles on Earth, repeating approximately once every 36 days.

See also

- Daylight saving time
- ISO 8601
- Jet lag
- List of time zone abbreviations
- List of time zones by country
- Lists of time zones
- Metric time
- Sunpendulum
- Time by country
- World clock

The control panel of the Time Zone Clock in front of Coventry Transport Museum.

References

[1] Latitude and Longitude of World Cities http://www.infoplease.com/ipa/A0001769.html
[3] Economics of Time Zones (http://bpp.wharton.upenn.edu/mawhite/Papers/TimeZones.pdf) PDF (1.89 MB)
[6] Quirico Filopanti (http://www.scienzagiovane.unibo.it/english/scientists/filopanti-1.html) from scienzagiovane, Bologna University, Italy.
[7] Gianluigi Parmeggiani (Osservatorio Astronomico di Bologna), The origin of time zones (http://www.iav.it/planetario/didastro/didastro/english.htm)
[8] The Astronomical Almanac 1983, US Government Printing Office (Washington) and Her Majesty's Stationery Office (London), page B4.
[9] Bowditch, Nathaniel. *American Practical Navigator*. Washington: Government Printing Office, 1925, 1939, 1975.
[10] Hill, John C., Thomas F. Utegaard, Gerard Riordan. *Dutton's Navigation and Piloting*. Annapolis: United States Naval Institute, 1958.
[11] Howse, Derek. *Greenwich Time and the Discovery of the Longitude*. Oxford: Oxford University Press, 1980. ISBN 0-19-215948-8.
[17] GetSystemTime function (Windows) (http://msdn.microsoft.com/en-us/library/windows/desktop/ms724390(v=vs.85).aspx)
[23] Chronos Date/Time Library (http://www.chronos-st.org/)

External links

- LEGAL TIME 2013 (http://www.itu.int/dms_pub/itu-t/opb/sp/T-SP-LT.1-2013-OAS-PDF-E.pdf)
- W3C Note Datetime (http://www.w3.org/TR/NOTE-datetime)
- US Official Time Clock (http://www.time.gov/timezone.cgi?UTC/s/0/java) Java-enabled clock to graphically display night and day around the globe.

Article Sources and Contributors

Hunters_Hollow,_Kentucky *Source*: http://en.wikipedia.org/w/index.php?title=Hunters_Hollow%2C_Kentucky *Contributors*: Acntx, CommonsDelinker, Hmains, Hushpuckena, John Cardinal, LlywelynII, Mithridates, Nyttend, Pearle, Ram-Man, Seth Ilys, Stevietheman, 1 anonymous edits

Bullitt_County,_Kentucky *Source*: http://en.wikipedia.org/w/index.php?title=Bullitt_County%2C_Kentucky *Contributors*: AXRL, Acntx, Ancjr, Antandrus, Brando03, Caknuck, CanisRufus, CapJ15, Censusdata, Chris the speller, Chris24, Coal town guy, D6, Dale Arnett, Danielba894, Debresser, Eastlaw, EngineerScotty, Flipperinu, GenSoup, Gobonobo, GrahamHardy, Hellmark, Hmains, Hushpuckena, Hyper Anthony, Ichabod, Indyjrg1762, Joshball, Kennethaw88, Leslie Mateus, Lightmouse, LouI, Lvillecards, Maxim, Mild Bill Hiccup, Mm40, Morgannxx63, Numbo3, Nyttend, Omnedon, Ost316, Patrickometry, Pearle, Ram-Man, Retired username, Robomanx, Ryuhaku, SGT141, Saxbryn, Sgt Pinback, Smallbones, Sportman122, Stevietheman, Susan12, TUF-KAT, Template namespace initialisation script, The Br3, Tvtonightokc, Whhalbert, WillC, Worldenc, 117 anonymous edits

United_States *Source*: http://en.wikipedia.org/w/index.php?title=United_States *Contributors*: !Silent, "Country" Bushrod Washington, (, (aeropagitica), *hollywood*, -Barry-, 041744, 067pansaol, 08OceanBeach SD, 08murtaghkc, 09LFIRTH, 0Bammer, 0alx0, 0ts0, 1.21 jigwatts, 1028, 10nn, 111Alleskönner, 119, 123Mike456Winston789, 1251thestrokes, 12835S, 1297, 149AFK, 1658, 16@r, 172, 18alex12, 19994best, 19starwars77, 1oddbins1, 1of3, 2005-03-04T23:39Z, 2005-03-09T21:41Z, 2005-04-08T13:18Z, 212etc, 23 editor, 23haveblue, 23prootie, 25 Cents FC, 2812, 28421u2232nfenfcenc, 2nd Piston Honda, 2t56, 3 years too many, 3232330, 400, 50imp, 5ptcalvinist, 621PWC, 66.81.31.xxx, 6r6r6muyr6uh, 7i5, 7mike5000, 84user, 8r1ck, 90, 987li, A Clown in the Dark, A Softer Answer, A State Of Trance, A Train, A bit iffy, A s williams, A-b-c-d-e-f, A-giau, A520, AAA765, ACD605, ACSE, ACW, ACupOfCoffee, AED, AEMoreira042281, AIRBUS, AKGhetto, AKMask, ALEXF971, AP1787, ARC Gritt, ARCoolguy, ARUenergy, ASDFGH, ASOTMKX, AStudent, AVM, AZCactus1, Aaker, Aalahazrat, Aang-kai, Aaporter 87, Aaron DT, Aaron Einstein, Aaron Schulz, Aaron mcd, Aaron7chicago, AaronS, AaronY, Abayarde, Abc is a number, Abce2, Abdalla A, AbeOlsen, Abecedare, Abilityfun, Abog, Abonazzi, Abtract, Abuchman1, Academic Challenger, Acalamari, Acanz, Accepttruth, Accurizer, Ace ETP, Acerperi, Acetic Acid, Acio83, Acjelen, Acmilan15, Acroterion, Adam Bishop, Adam Pamilih, Adam1213, AdamFirst, AdamJacobMuller, AdamRetchless, Adamski24, Adamsmo, Adamstraw99, Adanedhel21, Adashiel, Adavidb, Addicted04, Addisonn., Addit, Adelina and Hannah, Adhall, AdjustShift, Adjwilley, Adko110, Adnan bogi, Adolphus79, Adoniscik, Adpete, Adraeus, Adriaan, Adrian, Adrimaster3, AdultSwim, Aenar, Aeonx, Aerol, Aesopos, Aeyl13, Afil, Afinebalance, Afropuffs71, Against the current, Agdaman4life, Agtaz, Aherunar, Ahlabonde, Ahmad.ibn.as.Sayyid, Ahmadyk23, Ahodes1, Ahoerstemeier, Ahuskay, Aidepolcycne, Aido2002, Aim Here, Aiman abmajid, Aimjeff2, Airdish, Aislar, Aivazovsky, AjaxSmack, Ajwitte, Akamad, Akanemoto, AkvinRox, Akyoyo94, Al Villarruel, Al-Andalus, Al1encas1no, Alakasam, Alan Flynn, Alan J. Villarruel, Alan W, Alan.ca, AlanBarrett, AlanM1, Alandavidson, Alanhaggai, Alanscottwalker, Alansohn, Alarics, Albert0057, AlbertR, Albinostorm, Alcuin, Aldie, AlefZet, Alegoo92, Alephcat, Alex S, Alex43223, AlexJP, AlexNebraska, AlexOUK, Alexander Vince, Alexdragon, Alexignatiou, Alexihlo, Alexnovo, Alexsautographs, Alexwcovington, Alfiboy, Alhutch, Ali Karbassi, Ali a, Ali944rana, Aliazimi, Alientraveller, Alienus, Alilhappything, Alishakitty, Aliwaseem, Alkivar, All Male Action, All in, AlltimeJourneyFan, AllyUnion, AlmasterGM, Alodyne, AlphaEta, Alphabetagamnma, Altanar, Altenmann, AltiusBimm, Alvin Lee, Alvinrune, Amberrock, Amcfreely, AmericaIsNumberOne, AmericanXplorer13, Americanization, AmiDaniel, Amikeco, Aminatam, Amirb1, Amp873, Amren, An Siarach, AnOddName, Anakus, Anaraug, Anastrophe, Anativecantonesespeaker, Ancheta Wis, Andersmusician, Andkore, Andonic, Andre Engels, Andrea moro, Andrejj, AndresTM, Andrevan, Andrew Gray, Andrew Levine, Andrew Steller, Andrew c, Andrewbarber79, Andrewman327, Andreworkney, Andrewphelps, Andrewpmk, Andrewsthistle, Andris, Android79, Andy Marchbanks, Andy-ch, Andy120290, AndyBQ, AndyZ, Andyabides, Andyb215, Andychen86, Andypandy.UK, Andypham3000, Andyso, Anetheron589, Angel David, Angela, Angelastic, Angelboy, Angelo De La Paz, Angr, AngryParsley, Angryhobo16, Angusmclellan, Anhydrobiosis, Anilocra, Animum, Aninnymouse, Anirvan, Ankhet, Ankit jn, Ankitsingh83, Ann O'nyme, Anna Lincoln, Annama29, Annandale, Annasalia, AnnicAllus, Anon6666, AnonMoos, AnonyLog, Anonymity001, Anonymous editor, AnselmoLacerda, Anshumsaxena, Antandrus, Antarticstargate, Anthere, Anthony, Anthony Appleyard, Anthony717, Anti-Wuss, AntiBodies, AntiBones, AntiVanMan, Antimatter15, Anto475, Anton27, AntonioB, Antopi, Anubis2051, Anujill, Anywikiuser, Anzoooo, Aogouguo, Ap, Apatens, ApolloBoy, Apostrophe, Appleboy, Applemaster, ApprenticeFan, Aptdwn26, Apwvt, AquaD, Arad, Araiderplaya, Aranel, Aravind V R, Arbor to SJ, Arbuge, Arcadie, Arch dude, Archaelicos, Archaic, Archimedez, ArchonMeld, Arcimpulse, Arctic.gnome, AreJay, AreaControl, Argentsky, ArglebargleIV, AriGold, Aricci526, Aris Katsaris, Aristophanes68, Aristotle1990, Arkalochori, Armenia81, Armyrifle9, Arnon Chaffin, Arnoutf, Arpingstone, Arsenio, Arshadm, Art LaPella, Arthur Chevailier, Artifactblue, Artimbo, Artystyk386, Arwack, Arwel Parry, Arx Fortis, Aryavib, Asbestos, AscendedAnathema, Asderoff, Asdfjkl123, Asenai, AshAloy, Ashenai, Ashershow1, Ashnard, Ashrichards90, Ashton 29, Aspensti, AspiringScholar, Assyria 90, Asterix70, Astolfo Appleyard, Astonvilla91, AstroNomer, Astronautics, Astros4477, Astrotrain, Atchom, Atemperman, Athenean, AtilimGunesBaydin, Atlant, Atmans, Atomique, Atoric, AuburnPilot, Auburnirishman, Audacity, Aude, Aufheben, AuroraReloaded, Auroranorth, Ausimus, Austen Redman, Austen rules, Austin Hair, AustinKnight, Auswiger, Author782, AutomaticStrikeout, Autonova, Avala, AvalonXQ, Avargasm, Avaya1, Avenue, Avg, Avicennasis, Avidius, Avsrule1921, Awilcox1533, AxSkov, Axel Svensson, Axeman89, Aximill, Aydchery, Ayla, Ayrton Prost, AzaToth, Azmurath, B-watchmework, B. Fairbairn, B0at, B0mb3d, B1link82, B64, BD2412, BECASC, BG7, BGManofID, BGordon, BJCenters, BMF81, BRG, BSveen, Baa, Babablacksheephaveyouanywool, BabbaQ, Babbage, Babbby, Babygrand1, Babylicious2, Backslash Forwardslash, Backwalker, Badanagram, Badbe0, Bahn1k, Balderdash707, Ballboys10, Ballestrain, Bambuway, Bamford, Banders234, Bandity, Banes, Bansp, BanyanTree, Baplexed, Barbara Shack, Barbieblader, Barek, Barfooz, BaronLarf, Barrylb, Barryob, Bart133, BarzaniKurd16, Basbalfrk, Baseball Bugs, Baseballbaker23, Bash RF, Basilisk4u, Basohead, Bassbasketball03, Bassplr19, Bastique, Battem, Baxter 10, Bayonetblaha, Bazonka, Bbatsell, Bbenjoe, Bbert1994, Bbert1994cchs, Bbx, Bchalfin, Bdesham, Beach bum7, Beamathan, Beamer103, Bearingafish, Beastcuber, Bedford, Beef cakes, Beefy247, Beepoppab, Beetle1042000@yahoo.com, Beezer137, Before My Ken, Beh-nam, Behemoth98, Beland, Belchfire, Belligero, Beloki, Ben Arnold, Ben D., Ben Lunsford, Ben76266, BenFrantzDale, BenH, Benbristol, Benc, Benhocking, BenjaminTsai, Benji Powell, Benjiboi, Benkalt, Benlisquare, Benson85, Bentobias, Benwbrum, Berek Halfhand, Bet1973, Beta m, Betmeno, Bevo, Bevo74, Bgwhite, Bhadani, BhaiSaab, Bhamv, Bhe22, Bibbledebobblede, Bibbly Bob, Bidgee, Big Adamsky, BigBoyRubio, BigFishy, Bigairboarder121, Bigbear bh, Bigbluefish, Bigferdes, BiggKwell, Bigjust12345, Bigmak, Bigmanly, Bigredbrain, Bihco, Bilby, Bility, Bill McCloskey, Bill william compton, Bill-on-the-Hill, BillC, Billheller, Billim1, Billrnyc, BillyBoom, Billymac00, Binkhaldun, Biohazard930, Bipalabras, Bird1455, Birdmessenger, Birdofprey, Birthofthehive, Biruitorul, Bishonen, Bisteca, BitWarrior, Bjarki S, Bjmarfito, Bjoh249, Bjpremore, Bkell, Bkkeim2000, Bkonrad, Bky1701, Black Kite, Black and White, Black lyzard, BlackHades, Blackanthrope, Blackcell16, Blackmongoose, Blackshafen, Blainster, BlankVerse, Blarg2.0, Blastwizard, Blaylockjam10, Bleff, Blessthishouse, Blethering Scot, Blindman shady, BlisteringFreakachu, Blog Mav Rick, Blondace, Blood-Angel 9th Ranger, Blu Aardvark, Blue7545, BlueAg09, BlueSalo, Bluedog423, Bluedogtn, Bluegold, Bluemask, Bluemoose, Bluwiki123, BlytheG, Bmdavll, Bmgoau, Bmicomp, Bmitchelf, Bo, Bob A, Bob rulz, Bob schwartz, Bob swatsky, Bob121, Bobblehead, Bobblewik, Bobbob, Bobbum 47, Bobby Mouth of the South, BobbyAFC, BobbyJ, Bobbyi, Bobchalk, Bobdobbs1723, Bobdylan3589, Bobfreshwater, Bobisbob, Bobmack89x, Bobo The Ninja, Bobo192, Bobobobo, Bobozz, Bobtheuglyhobo, Bodnotbod, Bogdan, Bohemian tosspot, Bohuiginn, Boman1990, Bondegezou, Bonewah, Boneyard90, Bongwarrior, Bonked116, Bono983, BonsaiViking, Boogaryou, Boomshadow, Boothy443, Boraczek, Bornintheguz, Bosonic dressing, BostonRedSox, Bowei Huang, Bowei Huang 2, Boxbrown, BozMo, Bporter28, Br'er Rabbit, BradM, Bradifer, Bradwart, Braffit, Brainboy109, Brainhell, Brainstorm284, Branddobbe, Brandmeister, Brandmeister (old), Brandon, Brant4, Brat32, Bratsche, Bravada, Bread for the ducks, Breisblonde22, Brendan O'Shea, Brendanconway, BrendelSignature, Brennan626, Brenton, Brian Pearson, Brian the Editor, Brian0918, Brianga, Brickfield, Brickmack, Bridg1, Brighterorange, Brilliburger, Brion VIBBER, Brisvegas, Britannic1, Britcom, British legend, BritishWatcher, BritneySpearsAddict, Britsrule06, Broadside Perceptor, Brockert, Brodey, Broken Segue, BrokenSegue, Brokenfuse, Bronx Discount Liquor, Brossow, BrotherFlounder, Bruguiea, Brutaldeluxe, Brutannica, Bryan Derksen, BryanG, Bryant693, Brycehughes, Bssc81, Btipling, Buaidh, Buddha24, Buffalob07, Bugs5382, Bulba2036, Bull Market, Bulldog123, Bulldog73, Bullshark44, Bully25, Bunny-chan, Burningview, Burntsauce, Burroughsks88, Burto88, Bustoff, Butefraim22, Butros, Butwhatdoiknow, Buzity, Buzz-tardis, Bwinn, Bwithh, C. M. Harris, C.Fred, C.J. Griffin, C.Löser, CALR, CAMorris, CASportsFan, CDHgrün, CDN99, CJ, CJ Aikman, CJ King, CJK, CJLL Wright, CKBrown1000, CLW, CMSimons, CN3777, COF, COMPFUNK2, CORNELIUSSEON, CPacker, CQJ, CRouleau, CTF83!, Cacophony, Cactus.man, Cadiomals, Cafemusique, Cafzal, CalJW, Calcprgmer10011, CalicoCatLover, Califunya, Calliopejen1, Callmemrshowoff, Calltech, Calmypal, Calu2000, Calvin Lourdes He, Cam24, Cam275, Cambrasa, CambridgeBayWeather, Camenzind, Cameron, Cameron12345, Cameron12355, Camr, Camw, Can man138, Can't sleep, clown will eat me, Canada Jack, Canadaisawesome, Canadia, CanadianCaesar, CanadianLinuxUser, Canderson7, Caniago, CanisRufus, Canofmeat, Canterbury Tail, Cantus, Canuckian89, CapitalR, CapitalSasha, Capmo, CaptainNate, Captaincoach92, Carbonite, Cardiff123098, Cardinalwraith, CaribDigita, CarlosMarti123, Carnogaunt, CarolinianJeff, Carowinds, Carptrash, Carrp, Carthradge, Casper Gutman, Castletower, Castncoot, Catdude, Catswilltakeovertheworld1day, Causa sui, Cause of death, Caveman 07, Cbrett, Cbrown285, Cburnett, Ccacsmss, Cchow2, Ccs4ever, Cdc, Cdcon, Ce garcon, Cea801, Cecil722, Cecilgol, Cedric88, Cedrus-Libani, Celestianpower, Cello0wnsy0u, Cellurl, Celyndel, Censusdata, CentralTiger74, Centrx, Cenwin88lee, Cerealkiller13, Cerebrus13, CesarB, Ceyockey, Cfkane84, Cfquinlan, Cgoodell, Ch'marr, ChKa, Chadlupkes, Chanlyn, Channard, Channelone, Chanting Fox, ChaosSorcerer91, Chaosdruid, Chaosname, Charlie chaplin, CharlotteWebb, Charruss, ChaseJ589, Chason11, Chatori, Che829, Cheatcode2, Check two you, CheckPlus, Chedorlaomer, Cheese bombs, Cheese007, Cherkash, Chessman, ChesterTheWorm, Cheung1303, Cheungpat, Chiamus, Chic-ism, Chichui, Chickenofbristol, Chicopac, Chillum, Chindia (China-India), Chinju, Chinneeb, Chipmunk46, Chipmunkdavis, Chivista, Chobbs138, Chochem, Chochopk, Chocoforfriends, ChokeSlamFromHell, Cholmes75, Chongkian, Chris 73, Chris Mason, Chris cufc, Chris the Paleontologist, Chris-dotcom, Chris432, Chrisahn, Chriscoooley, Chrislk02, Chrismichener1, Chrisranders, Christian Historybuff, Christian List, Christinam, Christinewales, Christoph Braun, Christopher Kraus, Christopher Mulcahy, Christopher Parham, Christopher higham, Christophore, Chronus, Chrose, Chunky Rice, ChunkySoup, Chzz, CieloEstrellado, Cienade, Cieran 91, Ciroa, Cirt, Citation Needed, Citizen-of-wiki, Cityhighkid09, Civil Engineer III, Cje, Cjeiler, Ckatz, Clab6, Claidheamohmor, ClamDip, Clanget, Classic Fan, Clawson, Clayoquot, Cleared as filed, Clicketyclack, Clickpop, CliffC, Cloakdeath, Clorox, Closedmouth, CloudNine, Cloudblazer, Cloudreaver, Cma, Cmadler, Cmc0, Cmcginnis, Cmguy777, Coaster1983, Coasterlover1994, Cocacola125802, Cocoaguy, Cocoliras, Cocoma, Cod1337, Code Logic, Codingmonkey, Codu, Cofax48, Cohesion, Cold Light, ColdFusion650, ColdWarCharlie, Collect, Collier, Colonel Cow, Colonies Chris, Comayagua99, Cometstyles, Commander Keane, Commonbrick, CommonsDelinker, Compfreak7, Comput65, ComputerBuker, Computerwhiz171, Comrade Martin, Concordia, Coneslayer, Confiteordeo, Congolese fufu, Conk 9, Conman23456, Connelly, Conningcris, Connormah, Conor 33, Conquistador48, Conte di Cavour, Contemptiblepunx, Contra10, Conversion script, Conveyance, Conza1994, Cool Blue, Cool Stuff Is Cool, Cool-guy357, Cool4x411, Coolalec11, Coolcaesar, Cooldoug111, Coolhawks88, Coolio Grooverider, Cooliocheck1, Cooljuno411, Coolpeng123, Coolstorybrooo, Copperchair, Coraje, Cordless Larry, Cordova, Coredesat, Corgimaster, Corlen, Cormac Canales, Cornellrockey, Corriebertus, Corticopia, Corvun, Corwin8, Cosmic Latte, CosmologyProfessor, Costlab, Cougar Draven, Coulraphobic123, CoulterTM, Countersubject, Countmall, Courcelles, CovenantD, Coviepresb1647, Cow, Cowman109, Cpavlovski, Cprompt, Cptbuck, Cr0ssfire, Crad0010, Craigy144, Craky, Crash14, Crazy8jacky, Crazycomputers, Crazyparadigm, Crazytail2, Crazytales, Crd721, Credema, Creez34, Crevaner, Criptofcorbin, CristianoMacaluso, Critical.solvent, Crm18, CrnaGora, Crobzub, Crohall, Cronic 5085, Cros13, Crowbaaa, CrucifiedChrist, Crunchedfor6, Crusadeonilliteracy, Cruzian, Crxssi, Cryptic, CrypticBacon, CryptoDerk, Cryptosquid, Crzrussian, Cs-wolves, CsDix, Csigabi, Csurguine, Cthia, Ctryellowbird, Ctuavocado, Ctyankeeboy1638, Cuak's, CuboneKing, Cuchullain, Cuimalo, CumbiaDude, Cunado19, Curps, CurtisJohnson, Cutiewithabooty123, Cutrupi04, Cuye, Cvcdrk, Cverlo, Cyan, Cyanidethistles,

Cybercavalier, Cybercobra, Cyde, Cynical, Cynicism addict, Cyp, Cypher056E, Cyril Washbrook, Cyrilthemonkey, Cyrus111, D prime, D-Rock, D.M. from Ukraine, D.illah, D54b7ba, D6, DARTH SIDIOUS 2, DBaba, DC, DCB4W, DCGeist, DEW 7, DGJM, DHN, DJ Clayworth, DJHeartAttack, DL79OL, DLJessup, DMCer, DMellis, DTSX, DVD R W, DaGizza, Dabbler, Dablaze, Dabomb87, Dabrown, Dacoslett, Dacres, Daedalus969, Dah31, Dak is back, Dale Arnett, Dale-DCX, Daledavey, Dalek, Dalstadt, Damian Yerrick, Damicatz, Dammit, Dampe98, Dan Tarrau, Dan100, Dan56, DanGalligan, DanKeshet, DanRF, DanShearer, DancingPhilosopher, DangerousBob, Dangoodman, Daniel, Daniel J. Leivick, Daniel Morales, Daniel Quinlan, Daniel4sw, DanielCD, Danieldis47, Danielfarzan, Danieljo2013, Danielsavoiu, Danigoni, Dankie, Danny 17, Danoasis, Dans tes rêves, Dante2308, DanteAgusta, Dantesqueman, Danthemankhan, Dar-Ape, Daredude, Dargen, Dark Shikari, Dark glaive, Dark12334312, DarkAudit, Darkfrog24, Darkness Shines, Darknshadow, Darkwind, Darrenemerath23, Darth Panda, Darthbob100, Dasani, Dave Lars, Dave Runger, Dave.e.bondeson, DaveSDCali, Daveharvey83, Daven200520, Daveswagon, David H Braun (1964), David Johnson, David Kernow, David Levy, David s graff, David.Monniaux, DavidFarmbrough, DavidLevinson, Davidcannon, Daviddamoore, Davideus86, Davidgrazi7, Davidmcarson, Davidstrauss, Davidyoung, Davila, DavisGL, DavyJones14, Dawn Eastwood, Dawnseeker2000, Db099221, Dbabbitt, Dbachmann, Dbenbenn, Dbiehl, Dbtfz, Dc freethinker, Dcammz, Dcandeto, Dcflyer, Dday14044, Ddlamb, Ddssaa, Ddye, De Moriarty, DeLear2012, Deacon of Pndapetzim, Deanb, Deanos, Deathmelon92, Deathperson, Deathregis, Debate789, Deborahhosen, Debouch, Decumanus, Deeceevoice, Deep impact, Deepstratagem, Deeptrivia, Deetdeet, Defective 13, DefendEurope, DeifieD, Delaywaves, Deli nk, Delirium, Delj523, Delldot, Delta x, DeltaSpectre, Deltabeignet, Demomoke, Demon-lad, Demonix, Den fjättrade ankan, Denelson83, Denihilonihil, Dennis44, DennisAro, Deor, Der4, DerHexer, DerMeister, Deranged bulbasaur, Derek R Bullamore, Derek Ross, Dereks1x, Descendall, Destructoman, Deturtlemon1, Deus Ex, DevastatorIIC, Devil1337, Deville, DevinCook, Dewan357, Dewy9, Dfarmer, Dffgd, Dfrg.msc, Dgilman, Dgw, Dharmabum420, Dhn610, Dhruvhemmady, Diaments 7.0, Diberri, Dickrichards, Diderot's dreams, Differentgravy, DigbyDalton, Digitalme, Dillan.Murray, Dillon29, Dilweg, Dina, DinajGao, Dinopup, Dipankan001, Dipper2, Dirtjacket, Discordanian, Discospinster, Dismas, Disneyland Resort4, Dispenser, Dissident, Distal24, Diverse82, DividedByNegativeZero, Dizzyizzy, DjDark1 2009, Djinn112, Dkdantastic, Dlohcierekim, Dlwldud96, Dmartin969, Dmeek7750, Dmeranda, Dmerrill, Dnheff, Dniete97, DobbyD, Doc Quintana, Doc glasgow, DocKino, Docu, Doczilla, Dogma5, Dogs2000, Dolemite17, Dom188816, Domaleixo, Dominic, Dominics Fire, Domminico, Domthedude001, Donarreiskoffer, Donsiku, Donut2, DoomsDay349, Dooney 30, Doopokko, Doprendek, Dorfner, Doringo, Dostluk, DoubleBlue, Doubleodd, Doug Funny, Doug Johnson, Dougdorval, Doughboy, Dougweller, Doverspa, Download, Dp462090, Dpaajones, Dpark, Dpm12, Dposse, Dpotop, Dpr, Dr kapp 89, Dr.alf, DrVerlucci, Draconas, Draconix, DragonCurse, DragonFlySpirit, Dragonfir731, DragonflySixtyseven, Dragonsscout, Dragoon1025, Draicone, Drakky, Dralwik, Dran0259, Dranzini, Drcornwallis1, Drdyer, Dreaded Walrus, DreamGuy, Dreamafter, Dreish, Dres, Drestros power, Driftwoodzebulin, DriveMySol, Drmies, Drtjumper, Druid1090, Drumguy8800, Drummingboi, Dsaddsa, Dsfsdf, Dthdc4, Dtobias, Dtwiggy, Dubayuw, DubbaG, Ducker, Ducknish, Dude018219293, Dudewheresmypizz4, Dudtz, Dufekin, Dukakis1988, Duke Hazzard, Dumaka, Dunro, Durin, Durova, Durruti, DustFerret, Dustimagic, Dustintml, Dusty777, Dutchdrummer99, Dvdcdr, Dwilso, Dwrcan, DylanW, Dynam1te3, Dynamicknowledge28, Dysprosia, Dzulco, Dzzl, Długosz, E Pluribus Anthony, E Pluribus Anthony redux, E30e, EH74DK, ENeville, ESkog, Eagle4000, Earl Andrew, Earpol, Earth, Eastlaw, Easyid, Ebakunin, Ebolito, Eclecticology, Economiesofscale, EconomistBR, Ecopirate, Ectoplasmer, Ed Poor, EdBever, Edcolins, Edgar181, Edgars2007, Edgeshappy, Edgeweb, Editor18, Editor19841, Editore99, Edmundbaird, Edobrichoo, Edsanville, Eduardo Sellan III, Edward, Edward Z. Yang, Edward nz, EdwinHJ, Edwy, Effer, Eggeggeggegg, Egil, Eileen's Dream, Eink21, Einzig27, Eivindgh, Ek8, Ekcrbe, Ekrub-ntyh, El C, El Generico, El Slameron, El kulu ngile, ElBenevolente, ElTyrant, Elbbom, Elduncarno, Electionworld, Electromud, Electrosoccertux, Elfguy, Eli81993, EliasAlucard, Elitistnerd, Ella Plantagenet, EllenCT, Elliskev, Ellomate, Elockid, ElockidAlternate, Eloquence, Elpato, Elpaw, Eluchil, Eluchil404, Emanuel Kingsley, Emcee2k, Emigrant85, Emijrp, Emilson, Emmette Hernandez Coleman, Emorph, Emperors Harbinger, Empion, Emsko, Emx, Encephalon, Encheart, Enciclopediaenlinea, EncycloPetey, Energizer07, Energyfreezer, Enkrates, Enter Movie, Entirelybs, Enviroboy, Eoghan888, Eoin123, Eorde, Epbr123, Epeefleche, Epicadam, Epicgenius, Epiovesan, Equilibrium007, Erachima, Erfemf, Erginer, Eric outdoors, Ericamick, Ericci8996, Ericwirtanen, Erikrespo, Erudy, Erwetr, Esimal, Eskimo Boots, Esrever, Essaregee, Esvbie\eongvrklakjmatthew, Eswiki, Etaoin, Etcher, Ethan c.00, Ettrig, Eu.stefan, Eugen Simion 14, EugeneZelenko, Euphoriaq, Eurocopter, Eurofire, Euroleague, Evans1982, Evb-wiki, Evelhal, Evensteven1, Evercat, Everyking, Evhs00, Evil Monkey, Evil saltine, Eviloldman, Evolauxia, Excel23, Excimer3.141597, Existensorg, Exlibris, Extra999, Extremalnik, Eyu100, Ezhiki, Ezratrumpet, F. H. Black, F20ljc, FAH1223, FCStaehle, FDR, FESTORM 5, FEastman, FF2010, FFLaguna, FOARP, FSF-Rapier, FT2, Fabartus, Fabhcún, Fabrictramp, Facts707, Fadeddreams5, Faelomx, Faezdel, Fagnostic, Failofbeaner, Faisal 1918, Faithlessthewonderboy, Falconclaw5000, Falcorian, FallingGravity, Fame, Fangz, FargoWells, FarnhamJ, Fastily, Fatbug1, Fatla00, Fatpeople12345, Favonian, Fctk, FeanorStar7, Fearthesun, Featuresaltlakecity, Fedallah, Feinoha, FelisLeo, Fenice, Fennec, FeralitYzERO, Ferio252, Ferkelparade, Fernirm, Ferrariman, Feydey, Feyer, Feysaln, Fezzaga, Fhavenvt, Fibo1123581321, Fidelpmn2n, Fieldday-sunday, Fiet Nam, Fifty7, Filby, Filemon, FinkleDude, Finoman40000, Firebug, Firehose, Firemancvn72, Fishal, Fishiehelper2, Fitzy2280, Fixman, Fkarim, Flackson, Flag-Waving American Patriot, Flamanchti, Flamarande, Flamealchemist, Flashflash;, Flatterworld, Flaunted, Flexure, FloNight, Floaterfluss, Flockmeal, Flogiz, Flooey, Flosssock1, Fluent aphasia, Flurry, FlushinQwnzNyc, Fnfd, Folksong, Following specific instructions whispered by a mysterious cat, Fonduelazone, Fonzy, Football123213, Footballfan190, ForemanDomai, Former user 6, Forsyth.joel, Fortdj33, Fourthords, Fowlance, Fplay, FraDany, Fragsta, FrancisTyers, Francs2000, Frangibility, Frank, Frank0051, Frank94.kane, FrankCostanza, Frans Fowler, Frazzydee, Freak in the bunnysuit, Freakofnurture, Frecklefoot, Fred Bauder, Fredbauder, Frederico1234, Fredericoxxx, Fredlyfish4, Fredrik, Fredy.00, Freepenguin, Frencheigh, FreplySpang, FreshBulletTime, Freyr, Freyr35, Frichmon, Friedo, Frietjes, Frjlove, Frogsprog, Frood, Frostyservant, Frullic, Frustrated american progressive, Fshoutofdawater, Func, Fundamental metric tensor, Funnyhat, Funsides, Furdeen, Future2008, Futurist110, Fuzheado, Fuzzygenius, Fuzzyslob, Fvincent, Fvw, G Rose, G. Campbell, G.W., G2bambino, G8summit, GCarty, GD, GHe, GOP904, GRBerry, GRuban, GSTQ, Gabbe, Gabrielsimon, Gadfium, Gaff, Gagi, Gailbawlz, Gaius Cornelius, GaiusTimiusAwesomus, Gaiusknight, Gakon5, Galati, Galoubet, Galwhaa, Gamaliel, Gamer007, Gamer9832, GamerMod, Gamersedge, Ganesha1, Ganeshk, Ganrither, Garfieldt, Garfld986, Garkeith, GarnetRChaney, Garret Beaumain, Garretgwozdz13, Gary King, Garynine, Garytmlee, Gasolene, Gator1, Gatsby4life, Gauss, Gaylencrufts, Gaz, Gazpacho, Gbd1995, Gclinkscales, Gdo01, GearedBull, Ged was here91, GeeZee, Geeoharee, Gegik, Gekaap, Gene Nygaard, GeneralDuke, Genghiskhanviet, Gengis, Genius Chimpanzee, Gentgeen, Geo Swan, Geo0910, Geo247, Geofat, Geoff Plourde, GeoffreyVS, Geoking66, GeometryGirl, George, George415, GeorgeFormby1, Georgeadgdgdgdwngo, Georgia guy, Geothermal, Gerald Farinas, Gerg1031, Germanlight, Gershwinrb, Getcrunk, Getoar TX, Gfnb, Gggh, Gguit, Ghalas, Ghermit, Ghost1552, GhostPirate, Gianfranco, Giftlite, Gika, Gilderien, Gilgamesh, Gilhun, Gilliam, Gillydude, Gimferrer, Gimligimli1, Gimmetrow, GinaDana, Gingey933, Gintong Liwanag Ng Araw, GiovBag, Gjking, Gkmx, Gku, Glaucus, Glaxxo1, Glenn, Glenrqz, Globeism, Gman124, Gnaughton, Gnevin, Go229, Goalie1998, Gogo Dodo, Goheels36, GoingBatty, Gojoshg, Golbez, Gold Index, GoldRingChip, Goldblue21, Goldom, Gomarlins, Gomm, Good Olfactory, Good friend100, GoodDay, GoodSirJava, Goodandevil, Goodoldpolonius2, Goosehumper, Gop 24fan, Gopman1, Gordonfan, GorillazFanAdam, Goswamir14, Gr8opinionater, GraemeL, GraemeLeggett, GraemeMcRae, Graft, Graham87, GrahamTM, Gramaic, Grampion76, Grandin, Granpuff, Grant.Alpaugh, Graphic, Grapplequip, Graufbold, Green Giant, Green01, Greengreengreenred, Greenman, Greg96786, GregAsche, Gregcaletta, Gregorio, Greier, Greip, Grendelkhan, Gretzkyv99, Grey Shadow, Grey Wanderer, Greyhood, Grick, Grig541, Grimhelm, GrittyLobo441, Grjfiridkeokiski8j, Grm wnr, Grod028, Grop, Grosplant, Grosscha, Ground Zero, Grstain, Gruguuru, Grundle2600, Grunny, Grunt, Gsarwa, Gshave, Gsingh2300, Gsken, Gtuthtnfgurnfu, Guanaco, Guanxi, Gubbagubbaba, Guettarda, Gugilymugily, Guilherme Paula, Guitar 01, Gullit Torres, Gumbogumbo, Gundu1000, Gurch, Gurchzilla, Gurlpleezer101, Gurubrahma, Gustavb, Gutza, Guuzjo, Guy M, Gvhgu, Gwenham, Gwern, Gwernol, Gwickwire, Gwillhickers, Gwybedyn, Gypsum Miner, Gyre, H0ttamale2, H1478971, H5+R1A, HAHS 25, HIDECCHI001, HIZKIAH, HJ32, HTBrooks, HU Pilot, Habsfannova, Hackman911, Hadal, Hadrian89, Hadseys, Haemo, Hager jeff, Hahachuckles, Haham hanuka, HaireDunya, Hairy Dude, Haisthater, Hajor, Halberdier25, Halcionne, HalfShadow, Halibutt, Hall Monitor, Hallows AG, Hamedog, Hammer1980, Hans Dunkelberg, Happy-melon, HappyVR, Happyme22, Hardyplants, Harej, Hargrimm, Harkell, HarrietTubman, Harriganesh, Harro5, Harryboyles, Haruo, Harvestdancer, Hateless, Hateusa11, Haukurth, Haverberg, Havoc 8790, Hawaiian717, Hayden43, Haydenstone2, HazelGHC, Hazhk, Hcheney, Hchrishicks, Hcnorris, Hdsijhfoiuahfyusdaof, Hdt83, HeadCaptain, Headbanger756, Headlikeawhole, Heart Like a Wheel, HeartofaDog, Hearts of clear lake, Heat3000, Heaviestcat, Heby73, Hectorlf, Heegoop, HeikoEvermann, Heimdal, Heisencat, Hejsansvejsan, Hektossus, HeleneSylvie, Helios solaris, Helix84, Helixblue, Hellcat fighter, Hello14, Hello32020, Hello5959us, Helz0415, Henrik (old), Henry Delforn (old), Henry Flower, Henrygb, Hephaestos, Hequals2henry, Herdrick, HereToHelp, Hereforhomework2, Hernlund, Heroeswithmetaphors, Herp Derp, HerrGoebbles, Herunar, Herzie, Hexlor, Hfarmer, Hgspring, HiEverybodyHiDoctorNick, HiLo48, Highfields, Highvale, HillChris1234, Hillbillyholiday81, Himanis Das, Hipocrite, HippiedudeO, HippoOwl, HisForeShadow, Historian19, Historic Phoenix, Hiuby, Hk1992, HkCaGu, Hmains, Hmghosthost, Hobbentodd, Hobbesdude, Hobojoeandco, Hohum, Hoi jeon moo sool, Hojimachong, Hole62, Holledgeh, HollyAm, Holy Ganga, Homestarmy, Homeworld1031tx, HoodedMan, HooperBandP, Hoovernj, Hoping To Help, Hoponpop69, Hoppyh, HoratioVitero, Hordaland, Horses In The Sky, Hoshie, Hosterweis, Hot Stop, Hotelcalifornia1976, Hotshots2006, Hottentot, House1090, Househelps, Housesitter, Howard Alexander, Howcheng, Howdybob, Hseldon10, Hu, Huadpe, Huangdi, Hudude, Huduoroi, Huelenrite, Hugo Chavéz, Hugo999, Human diversity, Humanproject, Huphelmeyer, Hurricane111, Hurricanefloyd, Husond, Hut 8.5, HuuhManFui, Huzzahmaster018, Hydrogen Iodide, Hypathia, Hyperscope, Hystlen Caeth, Héous, I-2-d2, IANVS, IJA, ILguy06, ILuvAmerica, IMOW, IPodAddict181, IRelayer, ISpinksy, IZAK, Iakane49, IamBeastly20, IamYossarian, Iamthejake2000, Iamwisesun, Ian Dunster, Ian13, IanBailey, Iandiver, Ianhatch, Iankap99, Ibagli, Icairns, Iceberg3k, Icecradle, Iced Kola, Icedevil14, Id447, Idaltu, Idib, Ief, Ifrit 1780, Iggy402, Iggymwangi, Igilli, Ignacio Egea, IgnorantArmies, Igo4U, Ihatepotsmokinghippies, Ikanreed, Ikcyzrteip, Ike9898, Ikh, Ikyork, Ilikepie2221, Illegitimate Barrister, Illuminatiscott, Iluvmesodou, Ilya1166, Ilyanep, ImNotRichImStillLyin, Imaginary heroes, Imaninjapirate, Imbilly, Immunize, Impactplayer, ImperatorExercitus, Imperium Europeum, Improper Bostonian, Imzadi1979, In Defense of the Artist, InGearX, InShaneee, InTheRevolution2, Inanthropomorphism, Inchead, Indexheavy, Indopug, IndyLawSteve, Infidelis, InformationalAnarchist, Infrogmation, Inglok, Inhakito, Inimino, Inka 888, Innab, Insommia, Insuperable, Integralolrivative, Intelati, Intellibeing, Intelligentlove, Intelligentsium, Interfector, Internet quickie, Intimidated, Intmotinf, Introman, Intuitionz, InverseHypercube, Inwind, Ionutmortu, Iota, Iridescent, IrishHermit, Irishguy, Irishpyro, Irixman, Irrypride, Iruss14, Irviding11, Isawooty, Itai, Italia2006, ItalianGreyhound, Italianboy10, Italiano111, Italo, Iudaeus, Iuio, Ivankinsman, Ivannoriel, Ivejustbegun, Ivirivi00, Ixfd64, Izehar, J Clear, J Di, J intela, J. Daily, J. Finkelstein, J.delanoy, J3ff, J4bin, JACOB911, JAHL14, JB50000, JBH23, JCRules, JDooley, JEN9841, JForget, JG2K6, JGHowes, JHMM13, JHP, JHarrelson, JHunterJ, JIP, JJARichardson, JLMadrigal, JLaTondre, JMBZ-12, JMS Old Al, JPArnett, JPD, JPMcGrath, JPaestpreornJeolhlna, JStewart, JTBX, JTM, JVz, JW1805, JWSchmidt, JZZDD, JaGa, Jaberwocky6669, Jabolba, Jacda1313, Jacek Kendysz, Jacj, Jack.Hartford, JackDayton, JackLumber, Jackbirdsong, Jackdude101, Jacketpook, Jacksoncw, Jacksonsimon29, Jacky man Toronto, Jacob Lundberg, Jacob.clarke19617, Jacobah, Jacobolus, Jacottier, Jadabocho, Jafd88, Jak2, Jaked92, Jakro64, Jalada, Jalfredl, Jambalaya, Jamcib, James A. Stewart, James.d.mccollum, JamesMLane, JamesMadison, JamesMaxwell, JamesReyes, JamesTeterenko, Jamesday, Jamesmarkhetterley, Jamespeterka, Jamie C, Jamiemason, Jammy-dodger, Jamyskis, JanCeuleers, Jane tyra, Janechra, Janilson1995, Janviermichelle, JanvonBismarck, Jaranda, Jarble, Jared mckittrick, Jarheadigor, Jarjarbinks10, Jarkeld, JarlaxleArtemis, Jasgbro, Jason Potter, JasonAQuest, Jasoncward, Jasondmath, Jasonfitz, Jasper Deng, Jatkins, Java13690, Java7837, Jawsman2, Jaxad0127, Jaxbobby, Jaxl, Jay, Jay942942, Jayafrancis, Jaydee928, Jayemd, Jayhawk08, Jayron32, Jazz214, Jazz457, Jblounl, Jc900, Jcheckler, Jcm4, Jcmenal, Jcrocker, Jcw69, Jd027, Jdavidb, JdeJ, Jdforrester, Jdjoz0098, JeLuF, Jeancey, Jecar, Jecowa, Jedi master laybourn, Jeepday, Jef018, Jeff G., Jeff3000, JeffTL, Jeffgoin, Jeffrey S, Jeffz, Jenda H., Jengod, Jenks24, Jennavecia, Jennica, Jensboot, Jerahad, Jeremy221, JeremyA, Jeremygbyrne, Jeronimo, Jerry Jones, JerryFriedman, JerrySteal, Jerzy, Jess567, JesseRafe, JesseW, Jester2001, Jetlagorange, Jeyt5, Jez, Jezhotwells, Jezpuh, Jfarhou, Jfjkdc, Jfrancis, Jglees, Jgritz, Jguk, Jguk 2, Jh51681, Jhendin, Jhsounds, Jiang, Jibran1, Jihad!, Jim Douglas, Jim T, Jim1138, JimIrwin, JimWae, Jimbo D. Wales, Jimmy Kustes, Jimmy Slade, Jimmy butt, Jimmy-barnes, JimmySlade, Jimmyhead521, Jimothytrotter, JinJian, Jitse Niesen, Jivedevice, Jj137, Jj98, Jjacobucci, Jjb, Jjhcap99, Jkbena612, Jkfp2004, Jklin, Jkschroeder27, Jleon, Jlujan69, Jm51, Jmabel, Jman1974, Jmbrock, Jni, Joachimus, JoanneB, Jobe6, Jobroluver98, Jodafire, Joe Jarvis, JoeBlogsDord, JoeSperrazza, Joebieber, Joeblakesley, Joefromrandb, Joehaer, Joel27, Joelr31, Joelv40, Joesmoe8, Joesr55, Joeyramoney, Johan91, Johann Wolfgang, JohannVII, John, John Fader, John Price, John Quincy Adding Machine, John Smith M.D., Ph.D, John Torn, John-1107, JohnDBuell, JohnOwens, JohnRuskin, JohnWittle, Johndoe85839, Johnian144, Johnleemk,

Johnmc318, Johnmike81, JohnnyMrNinja, Johnowenlangham, Johnstone, Johntex, Johntinker, Jojhutton, Joker123192, Jokeroler, Jon Doh, JonMoore, Jonarooni, Jonas Viper, JonasanRat7, Jonathan Drain, Jonathan-Morris711, JonathanFreed, Jonathunder, Jondel, JonesMama, Jongpil Yun, Jonnie d smith, Jonny Nixon, Jonsafari, Jonyyeh, Jooler, Joolz, Joowwww, Jophus00, Jorcoga, Jordanian forever, Jordz, JorgePeixoto, Jorgezoltan, Jose77, Joseph Buchanan, Joseph Solis in Australia, Josh Parris, Joshdboz, Joshua Scott, Joshua368, Joshurtree, Jossi, Josteinaj, Jowe189, Joy, Joylifen, Joyous!, João Neves, Jpers36, Jpgordon, Jprg1966, Jprulestheworld, Jrboi, Jrdioko, Jredmond, Jrkarp, Jrkso, Jrn, Jrobin08, Jrockley, Jros83, Jrt989, Jrtayloriv, Js00902, Jsoltman, Jtdirl, Jtkiefer, Jtokarsbr, Jts6293, JuJube, Juantay, Judgement, Judgesurreal777, Judygreenberg, Julain Barbarosa, JulieADriver, Jumbuck, Jumple, Jumpythehat, Junes, Jupiter Green, Juppiter, Jusdafax, Jusjih, Jusme1, Just a member, Justanotherguitarist, Jwissick, Jwking, K. Lastochka, K157, K50 Dude, KBtoys09, KConWiki, KF, KFan II, KGBarnett, KGasso, KL22, KS-U92, KTC, Kaasje, Kablammo, Kadelia, Kafziel, Kaio-ken x10, Kaisershatner, Kajal4, Kajmal, Kaldari, Kalsermar, Kampfers, Kangy, Karam.Anthony.K, Kardrak, Karimarie, Karl, Karl Dickman, Karukera, Kasyapa, Katalaveno, Katefan0, Kathmandu2007, Katydidit, Kaveh, Kazikame, Kazu-kun, Kazvorpal, Kbdank71, Kbolino, Kedar63, Kedi the tramp, Keebler71, Keegan, Keegster5, Keelm, Keilana, Keith D, Keith H., Keithd, Kellen, Kelly Martin, Kelstonian, Kelvinc, Kember, Kemet, Kemiv, Kemmabrant, Ken Seh, Kenaldinho10, Kennard2, Kenndo, Kennethduncan, KensterFox, Keon Milton, Kerrysfrench, Kertof, Kesla, Kesselring88, Kethinov, Kev2310, Kevin B12, Kevin Myers, Kevin Taylor, Kevin12xd, Kevinmon, Kevinp2, Kevjumba, Kevs, Kewp, Kg vasilev, Kgrad, Kgrr, Khajidha, Khalid, Khanada, Khatores, Khatru2, Khazar2, Kiddster93, Kier07, Kieranandersn, Kievaughn, Killdevil, Killer Swath, Kimon, Kinaro, King Dracula, King Gemini, King of Hearts, King wiston, KingBob1, KingTT, Kinggeedorah, Kingpin13, Kingturtle, Kinneyboy90, KinturkMan, Kirby of Doom, Kirchherr, KirchnerE07, Kire 1123, Kirill Lokshin, Kirkieboy, Kishorekumar 62, Kitch, Kitsunegami, Kjaer, Kkm010, Klaam, Klafubra, Klander Brigade, Klaser, Kliituu, Klonimus, Klr390, Kmacsimus, Kman543210, Kman665, Kmanblue, Kmaster, Knife2sharp, Knightshield, KnightxxArrow, Knorpel, KnowledgeOfSelf, Knowsitallnot, Knoxbury, Knutux, Koavf, Kochamanita, Kogsquinge, Konstantin, Korora, Kotjze, Kotniski, Koyaanis Qatsi, Kozuch, Kpjas, Kr1st1deejay97, Kraalg, Krakaet, KrakatoaKatie, Kralizec!, Krator, Krellis, Krich, Krismanx, Kristian Vangen, Krizpykremekid, Krupo, Kschwerdt514, Ktsquare, Kttreasure, Kude90, Kudzu1, Kueldude, Kukini, Kulikovsky, Kumioko (renamed), Kunaygi, Kungfuadam, Kurieeto, Kurt, Kurtle, Kuru, Kurykh, Kusma, Kuzco Mii, Kuzmaxster, Kwame1234, Kwamikagami, Kwijiborjt, Kwirky88, Kwizzard, Kxcd, Kyh87, Kyle Barbour, Kyle J Moore, Kyle sb, Kyle1278, Kyle4321, Kzzl, Kızıl Şaman, L33th4x0rguy, L3p3r, LGagnon, LIbertyInSpace, LOL, LaMenta3, Labby24, Labeledstickmen, Labz, Lachatdelarue, Lacrimosus, Lactical, Ladb2000, LadnaV, Laguna72, Lakshmix, Lala0000, Lalein91, Lamp coin, Lampoonsvacation, Lan56, Lanboney, Langston, Laomei, Lapsus Linguae, Lardy123, Lasindi, Last king of Frisia, LaszloWalrus, Latics, Latitude0116, Lauboo, Laubz83, Laucheehung, Laudaka, Lauren, Laurinavicius, Lawe, Lawrencekhoo, Lawyer2b, Laxbrain0110, Lazyandy123, Lcarscad, LeAviateur537, Lear 21, LeaveSleaves, Lec domicilarias, Lectonar, LedRush, Ledzeppelin321295, Lee Daniel Crocker, Leeannedy, LegitimateAndEvenCompelling, Lego872, Legoboy2000, Legokid, Legoktm, Lehi, Leif, Leithp, Lemonaderox, Lenoxus, Lenzar, Leodj1992, Leon7, Leonard G., Leroyinc, Leszek Jańczuk, Leuko, Levineps, Lewisgs, Lexor, Lfdder, Lfgnyc, Lfh, Lhademmor, Liber, Liberalcynic, Liberi Fatali37, Liechten11, LifeStar, Liftarn, Lightdarkness, Lightlowemon, Lightmouse, Lightningfro, Ligulem, Lijealso, Likeminas, Lilac Soul, Lilpinoy 82, Lima bean of the north, Limideen, LincolnSt, Lindberg, Lindenhurst Liberty, Ling.Nut3, Linuxbeak, Lipton1995, Lisa517, Lisazipper, LittleDan, LittleRoughRhinestone, Littlelance, Livajo, Living001, LizardWizard, Lizzie Harrison, Ljriccio9j, Llakais, Llec, Lloyd.smith, LlywelynII, Llywrch, Lmusielak, Lneal001, Locaracle, Lockesdonkey, Logan, Lokibolp, Lolnoraj, Londomollari42, LonelyMarble, LongBay, Longhorns4life89, Longwalkshortpier, Loodog, Looper5920, Looxix, Lord Emsworth, Lord Geznikor, Lord Moostyer, Lord Terminus, Lord Voldemort, Lord Wikipedian of the Wiki, Lord alexandre, LordHarrison, LordSimonofShropshire, LordSnooze, Lordflamel, Lordvaluemart, LoreleiLynn, Lorenzop, Lorenzosabbadini, Loser111111111122222, Lotje, Lou1986, LouI, Louis Waweru, Lousyd, Louturks, LovesMacs, Lowellian, Lowreeder, Lowride, LoyolaDude, Lrusso99, LtWinters, LuGiADude, LuK3, Luby, Lucas Duke, LudicrousTripe, Ludvikus, Luigi5256, Luiscolorado, Luisztdt, Luk, Luke334, Luke56, LukeM212, Lukemadness, Luna Santin, Lupin, Lupita°, Lupo, Lupus Bellator, Lussier, Luxor99, Lycurgus, LynnMeilleur, Lynntoniolondon, M a s, M2MM4M, M5891, MAG Fry, MBisanz, MC MasterChef, MCRTGP, MDCore, MECU, MER-C, MJCdetroit, MJDTed, MJEH, MJK, MKers06, MONGO, MPS, MWaller, MWatson15, MZMcBride, Mabbass, Mac Davis, MacGyverMagic, Macfan93, Macg4cubeboy, Macintosh User, Macjaeh, Mackay 86, Mackensen, Mackoy20, Macphisto12, Madbiggie00, Madcoverboy, Maddiekate, Madman445, Madskillz225, Maelnuneb, Maestro25, Magicman710, Magister Mathematicae, Magister Scienta, Magnet paper, Magog the Ogre, Maheshvenk, Maijinsan, Mailer diablo, Mailman9, Mais oui!, Majoreditor, Majorly, Makaristos, MakeChooChooGoNow, Malarious, Malcolm, Malhonen, Malo, Maltmomma, Mana Excalibur, Mandey7, Manicsurge, Manix, Manop, Mansmokingacigar, MantisEars, Manutdglory, MarSch, Marcfromns, Marcika, Marco Neves, Marco polo, MarcoTolo, Marcoescobar, Marcopolo7890, MarcusMaximus, Marcusliou5, Marcworca, Mareino, Marek69, Marielleh, Marine(uc), Mark, Mark Antony666, Mark Miller, Mark Renier, Mark Sheridan, Mark83, MarkDalit, MarkDeyarmond, MarkSweep, Markpravda, Marmote, Marnanel, Marrovi, MarsInSVG, MarsRover, Marshall Williams2, Marshmallow00, Marshman, Marskell, Martarius, Marth2k, Martin S Taylor, Martin451, MartinHarper, Martinense, Martinp23, Martinwilke1980, Martious, Martyman, Martynrss2, Marvolo Gaunt, Marymer, Marysunshine, Mashboy, Mason.Jones, Massimo Catarinella, Master Jay, Master and Commander, Master of Puppets, Master of the Orichalcos, Mastercomputerpro9999, Masterjamie, Masterpiece2000, Matchgame12345, Mateo SA, Materialscientist, Materic123, MathewDill, Mathgod333, Mathpianist93, Matsaball, Matt 8, Matt Crypto, Matt Gies, Matt Hales RDFC@hotmail.com, Matt Heard, MattPDG, Mattabesta, Matthewprc, Mattkidd12, Mattt296, Mattworld, Matty 1988, Matty.007, Mattyh190, Mattyhar, MaulYoda, Mav, Max rspct, Max(is not)well, Max20020012002, Maxatl, Maximus Rex, Maximuspower, Maxkoyama, Maxkramer, Maxmc, Maxpower37, Maxsht9, MaxxFordham, Mayumashu, Mb93j, Mblumber, McBenjamin, McDogm, Mcaleppo86, Mcattell, Mceder, Mchcopl, Mcimo88, Mckyj57, Mcmoor, Mcsweet, MeThOd, MeasureIT, Meco, Medevila, Medo4, Medo9, Medvedev40, Meegar, Meelar, Meeptastic, Meeve, MegX, Megaandroid30, Megahmad, Megan 189, Megapixie, Megawattbulbman, Mehicdino, Melonkelon, Melsaran, Member, Menchi, Mendel, Merbabu, Merlot70, Mermaid from the Baltic Sea, Merovingian, Mesgul82, Messiisking, Metafisyx, Metehan1992, Mets501, Metz2000, Meursault2004, Mevlüt Kılıç, Mexihcatlacatl, Mfwnoface, Mghabmw, MghtyRoach, Mgiganteus1, Mh10190, Mhesselb, Mhking, MiKellie, Michael Essmeyer, Michael Glass, Michael Hardy, Michael Patrick, Michael of the World, Michael-Billa, Michael.Urban, Michael2-, MichaelBillington, Michaelenandry, Michaelpkk, Michaelwuzthere, Mickey gfss2007, Mickeykozzi1, Mickeymickey11, MicronationKing, Midlothian09, Midnight Green, Mightymights, Mihas, Mika293, Mikael Häggström, Mike Christie, Mike Halterman, Mike Pehipol, Mike Rosoft, Mike nofx, MikeBravo, MikeLynch, Mikefzhu, Mikehawk, Miken32, Mikeroodeus, Mikestern2, Mikeybadass05, Mikhail Ryazanov, Mikhailov Kusserow, Mikinator, Mikpass1, MilFlyboy, Mileage, Milk of magnesia, Milk the cows, Milkmandan, Millahnna, Million Moments, Millsdavid, MindscapesGraphicDesign, Minesweeper, Minesweeper.007, MiniZiper, Minjamonkeyw00t, Minna Sora no Shita, MinnesotanConfederacy, Miranda, Mirrorkirby9062, Mirv, MisfitToys, Miskin, Miss tylie, Miss-jessie-gal, Missionary, Missmarple, Missouri Rebel, Mistakefinder, Mister Alcohol, Misterrick, Mistertheman, Misza13, Mitaphane, Mitchumch, Mitsukai, Mitth'raw'nuruodo, Mixcoatl, Miyokan, Mizuphd, Mjf3719, Mkweise, Mlazik, Mm40, Mmarch31, Mmavipc, Mmx1, Mmxx, Mnlira013, Mnmazur, Mo0, Moby-Dick3000, Moderate2008, Modest Genius, Modulatum, Moe Epsilon, Mogden, Mogism, Momentary, Momo san, Momus, Mon amour est avec la France, Moncrief, Mongoletsi, Monkeyblue, Monkeyman, Monkeywrenchjim, Mono, Monobi, Monocletophat123, Monsieur Fou, Monsterhead150, Montrealais, Montyofarabia, Moondog88, Moonraker0022, Mooseberry, Mordien, Moreau36, Morgankevinj huggle, Morhange, Morphh, Morriswa, Morton devonshire, Morwen, Mosca, Mosemamenti, MosheA, Mostgiganticlyingmouthofalltime, Mosthauntedjp, Mostly Rainy, Motthoangwehuong, Mouloud47, Moustan, Moverton, Moxy, Mozartmuzic, Mprudhom, Mpscan, Mr Adequate, Mr Stephen, Mr Taz, Mr Twix, Mr. D Vega, Mr. Lama, Mr. bwn, Mr.Do!, MrD9, MrDrew508, MrGRA, MrJanitor1, MrMick, MrRadioGuy, Mrbelvedereposter, Mrbrown, Mrcolj, Mrdthree, Mrjpb104, Mrknowitman, Mrld, Mrmaroon25, Mrtobacco, Mrwojo, Mrzaius, Mschiffler, Msludwick, Msundqvist, MtnBrook, Mufka, Mulad, Mumble45, Munulu999, Muraad kahn, Murali nmv, Murchy, Murdocc, MurphysLawn, Musdan77, Mushroom, Musicalantonio, Musiphil, MuskokaON, Musse-kloge, Mustanglover, Muya, Mwanner, Mx3, Mxcatania, Mxn, Myanw, Mydreamistofly, Mygeneration01, Myleslong, Mylittlezach, Myriad, Mystaker1, Mystify85JEC, Mário, Mäximus Rex, N-edits, N-true, N.M.Sheedy, N00bschoolbus, N244888, N5iln, N734LQ, NBS525, NByz, NDSteve10, NGerda, NHRHS2010, NJOrlando, NJRobbie, NJguy281, NKT45, NSLE, NSR, NW036, NWOG, NYCJosh, NYKevin, Nach0king, Nachomania, Nadavspi, Naddy, Nader85021, Nafield, Nakon, Nakul Dhoot, Nappyrootslistener, Narco, Nareek, Naryathegreat, Nascar1996, Nat, Nat Krause, Nat682, Nate1012, NateXMS, Natgoo, Nathan8225, Nathanalex, Nathanbolt, Nathanshinsen, NativeForeigner, Nattuvijay76, Natty4bumpo, NatusRoma, Naus, Nautilator, Naval Scene, NawlinWiki, Nbatra, Nblc, Nbrzycki91, Nd24, Nduhn101, Neal Finne, Nebulousity, Nehrams2020, Neilbeach, Nelandquinten, Nelfer, Nelro, Nelsondavidd, Neo233, NeoLamagra, Neoballmon, NeonNight, Neondiscogod, Neorunner, Neovu79, Nerdalert101, NerdyScienceDude, Nerothanes, Netbenefit, Netesq, Netkinetic, Netoholic, Netscott, Neuroscientist, Neutrality, Nevada, Nevilley, New England, NewEnglandYankee, Newstormer, Nezbie, Nhakimi, Niceguyedc, Niceley, Nichalp, Nichofas, Nicholas F, Nicholas Tan, Nick, Nick.mon, Nick123, Nick4gwen, NickMartin, Nickptar, Nickyaberdeen, Nico, Nicolas Barbier, Nicolewasherexx, Nicolharper, Nidam, Niemti, Nige, Nigeria01, Nighend, Nigholith, NightDragon, Nightscream, Nihonjoe, Nika, Nikkimaria, Nikodemos, Nikofeelan, Nikola Smolenski, Nikoniko, Nikostar, Nikwad244, Nil Einne, Nil105994, Nimbley6, Nine999999999s, Nipplemaster5000, Nippoo, Nirvana888, Nishant12, Nishishei, Nissi Kim, Niteowlneils, Nitharsank, Nitya Dharma, Nivix, Njoedits, Nkrosse, Nmpenguin, No Guru, No toco at aol dot com, Noleander, Nomad, Nomist, Nonenmac, Noozgroop, NorCal764, NoraBG, Northamerica1000, Northmeister, Northwales4u, Northwest, NortyNort, Norwikian, Nosimplehiway, Nosskyline, Not a dog, NotAbel, Nothlit, Nottheking, NovaNardis, Novidmarana, Novis-M, Nowhither, Nrafter530, Nrana99, Nroscha, Nscheffey, Nskrill, NuKkEm, NubKnacker, Nuberger13, NuclearVacuum, NuclearWarfare, Nufy8, Nuge, Nukebros, Numberoneconti, Numberonesurvivor, Numerouno13, Nunh-huh, Nvw, Nxxos, Nydas, Nzzl, OBrienMadHouse, ONEder Boy, OOODDD, Obaidz96, Obhave, Obi-WanKenobi-2005, Obli, Oblivious, Occono, Octane, Oda Mari, Offenbach, Offrecord09, Ohconfucius, OhitstragikX, Ohmygodeven, Ohnoitsjamie, Okiefromokla, OlEnglish, Olaf Eriksson, Old Guard, Old Moonraker, Old Right, OldRight, OldRightist, Oldag07, OldakQuill, Oldmankdude, Ole10589, OleMaster, Oleg Alexandrov, Olegwiki, Olivier, Ollieegg, Olorin28, Omarlegend, Omarmorek, Omega9380, Omicronpersei8, Ommnomnomgulp, Omnibus, Once in a Blue Moon, One, One man orgy, OneDoubleO, OneEuropeanHeart, OnePt618, Onetwo1, Onixz100, Onopearls, Onorem, Oo64eva, Oobopshark, Opaldragon1, Opelio, Openthedoor754, Operation Spooner, Optim, OptimumPx, Optional Toaster, OrbitOne, Ordealbyfire, OreL.D, OregonD00d, Oreo Priest, Oreotoast, Orthorhombic, Oscarthecat, Oshwah, Oskart, Ost316, Otashiro, Otebig, Otolemur crassicaudatus, OuroborosCobra, Outback the koala, Outoforderuk, Ovcp1, Owen, OwenBlacker, OwenX, Oxboomer, Oxford Comma, Ozi, P.W.Lutherson, P0per, PAWiki, PBS, PCock, PDH, PFHLai, PL290, PRRfan, PRSteve27, PS2pcGAMER, PZFUN, Pabejar, Paco8191, Pah246, Pahistorian, PainMan, Paine Ellsworth, Paix Pan, Pajz, Pak21, Pakaran, Paleo dragon, Paleoblues, Pamri, Panfakes, Panserbjorn, Pantherskin, Papasmurf0810, Papercrab, Para, Parachnidox, Paradise7, Parallel or Together?, Paraside, Parliamentcity, Parmadil, Parthasarathy B, ParvatiBai, Pass a Method, Pastorsson93, Patapsco913, Patioco, Patka, Patman2648, Patrick, Patrick19, Patrickneil, PatrikR, Patsson77, Pau07, Paul A, Paul August, Paul Tyrrell, Paul.h, Paul2.0, PaulHanson, Paulinho28, Pauljeffersonks, Paulmcguane, Paulmec, Paulsprecker, Pax:Vobiscum, Pb30, Pbedford, Pcman22594, Pcpcpc, Pcstico, Peak, Pearle, Pecher, Pedant, Pedantic Turkey, Peeweeteehee, Pegasus1138, Pekin Republican, Peloneous, Pendragon5, Pengkeu, Penny1020, Penser, Penshurstpride, Penubag, Pepesia, Pepper, Pepsi132, Pepsidrinka, Per Honor et Gloria, Peregrine Fisher, Perfect Editor, Perfect77, PerfectStorm, Perfectionaintperfect, Perkinator 3000, Permethius, Perseus, Son of Zeus, Persian Poet Gal, Peruvianllama, Pestcamel44, PeteJayhawk, Peter, Peter L, Peter.C, PeterK, Peterale4444, Peterlin, Pethr, Petrb, Petrux, Pewwer42, Peyman Ghasemi, Pfortuny, Pgan002, Pgecaj, Pgk, Ph.D.Nikki, PhJ, PhaedoHD, Phamdangkhoa2410, Phantomsteve, Pharaoh of the Wizards, Pharos, Phe, Phil Boswell, Phil5329, Philanthropy, Philip Stevens, Philip Trueman, Philip200291, Phillies1fan777, Philliesfan035, Phillip Fung, Phillip J, Philpill691, Phinnaeus, Phoebe, Phoenix2, Phoenix7777, Photo me2004, Phrankle, Phrasia, Phxcesar, PhysicsPat, Pi399, PiMaster3, Picapica, Pidgeontoes, Pierco, PierreAbbat, Pigman, Pilko182, Pillow2011, Pilotguy, PinchasC, Pine, Pinotgris, PinoyWikipedia, Pioelad, Piotrus, Piss420, Pizzadeliveryboy, Pk10534, Pkpat2011, Plainsong, Plapre09, Plastic, Plasticbadge, Plusgood, Pm06420, Pmackarate, Pman31014, Pmanderson, Pnatt, PoccilScript, Poetaris, Poilil, PokeHomsar, PokeYourHeadOff, Pol098, Polaron, Polarscribe, Polisci, Politepunk, Political hack, Politis, Polylepsis, PonileExpress, Poojean, Poopymatt, Poor Yorick, Poppa Bear, Porkman385, Porsche997SBS, Portillo, Postco, Postdlf, PotentialDanger, Potnpan123, Pperos, Prasenberg, Pratheepps, Premeditated Chaos, Preschooler.at.heart, Preslethe, PrestonH, Pretty Green, Preuninger, Primalchaos, Primarypenguin, PrinceForte, Prision, Pristino, Prjrv787, Prodego, Producercunningham, Prospect1, Proteus, ProudIrishAspie, Pryde 01, Pschemp, Pseudo-Richard, PseudoSudo, Psy guy, Pubber, Publicus, Puertorico1, Pulmonaryarchery75, Pupitar13, Purodha, Pushbreak, Pwilk,

Pwjb, Pwnage8, PxT, Pyroclastic, Pyroechidna1, Pyukman, Pz821, Pé de Chinelo, Q Ramona, Qasaqsuyu, Qmwne235, Qqqqqq, Qrc2006, Quac, Quadel, Quadell, Quadrochosis, Quaker24, Quako5, Quanticles, Quasipalm, Quazywabbit, Quberoot, Query, Questchest, Quijebo, Quill, Quinn C56, Quinnculver, Quite vivid blur, Quixoto, Quizimodo, Quizimodo (usurped), Quota, Qviri, Qwe, Qwerasdfzxcv1234, Qwfp, Qxz, R'son-W, R-41, R-Joe, R0xorz, R9tgokunks, RA0808, RG2, RIPSAW1986, RJII, RJN, RMHED, RPIRED, RPlunk, RPlunk2853, RS085, RTG, RVD, RVJ, RWhite, RZ heretic, RaCha'ar, Rab777hp, Rachaelho50, Radarsnobles, RadicalBender, Radicalsubversiv2, RadioBroadcast, RadioKirk, Raegis, RafaelG, Rageinacage, Ragib, Rahlgd, Rahuldharmani, Rail88, Railer 893, RainR, Rainpat, Raistuumum, Raja Hussain, RajaNeela1993, Ral315, Ralph3000, Rama, Rama's Arrow, Ramanglass544, Ramazan!, Ran, RanEagle, Ranainam5343, RanchoRosco, RandalSchwartz, Rande M Sefowt, Random contributor, Randomblue, Randombob12345, Randomengine, Randy Johnston, Randy Kryn, Rangermike, Rangerwave, Rangoon11, Ranveig, Raoul-Duke, Raprchju, Rarelibra, Ratemonth, RattleMan, Raul654, Ravedave, Ravelnine, Ravenhull, RaviC, Ravikiran r, Ravlo2004, Raybob95, Rayc, Raymond arritt, Rayoflight278, Rb2911, Rbankson, Rcharding, Rctbone, Rd232, Rdrgz93, Rdsmith4, RebaFan1996, Rebellion101, Reboot, Reck345, Red Slash, Red rawker, Red65, RedChord, RedWordSmith, Redd Baron1, Redd Foxx 1991, Reddi, Redeem, Redgecko5, Redkloud, Rednaxela, Redwolf24, Reeces904, Reedy, Reelcheeper, Refdoc, Regancy42, Regionalsimp, Reid A., Reid1867, Reillyg, Reimelt, Reinthal, Reisio, Rejs12345, RekishiEJ, Reliancepowercoin, Rembo74, Remiel, Renegadeviking, Renorei, Rentastrawberry, Renzo92x, Republicanbetter, Republikaner, Reticulum, Retired username, Retireduser1111, Retrospector87, Rev.Ision, Revmachine21, Revolución, Rex Germanus, RexImperium, RexNL, Rextrim, Reywas92, Rfc1394, Rgrigal, Rhaegar, Rhatsa26X, Rhobite, Rholton, Ric36, Rich Farmbrough, Richard Allen, Richard Blatant, Richard David Ramsey, Richard0612, RichardDavila, RichardF, Richie123098, Rick Block, RickK, Rickipedia127, Rickney, Rickvaughn, Ricky@36, RickyTheBest, Rickyrab, Riconoen, Rifecu, RightCowLeftCoast, Rillian, Rimbaud, Riphamilton, RiseAbove, Risker, Ritchy, Ritimakovej, Rittmeister, Rivertorch, Rjairam, Rjd0060, Rjd2 Deadringer, Rje, Rjensen, Rjm656s, Rjwilmsi, Rklawton, Rkt2312, Rlevse, Rlm0710, Rlundi86, Rmerlos, Rmhermen, Rmhs15, Rmky87, Rob 06-07, Rob Shepard, RobTranter, Robbiemt, Robbo chappers, Robbro7, Robchurch, Robdurbar, Robert K S, Robert.Mcleod, Robert25, RobertG, RobertHuaXia, Robertkeller, Robertson-Glasgow, Robertvan1, Robfergusonjr, Robin63, Robster1983, Rocafella, Rocastelo, Rocawear1994, Rochabm, Rockentui, Rockfang, Rockstone35, Rockyobody, RoddyYoung, Rodrigogomespaixao, Rofl, Roger Arguile, Roger Gianni, RogerK, Rogered, Rogertudor, Rogueagent123, Rohawn, Roke, Roleplayer, Rollercoasterlord, Rollo44, Rolo75, Romanm, Rompe, RonH, Ronaldmolina20, Ronster21, Ronthewizard1994, Roosevelt Churchill, Rory096, Rory2259, Ros24s, Rose Garden, Rossenglish, Rosshterry, Rossnixon, Rotten, RoundDaWorld, Route11, RoverTheBendInSussex, Rox Da Box, Roxi2, Roxorzursoxorz, Roy Brumback, Roy da Vinci, RoyBoy, Royboycrashfan, Rreagan007, Rrhain, Rrie, Rrius, Rror, Rrouse, Rsm99833, Rstrug, Rto and buddy, Rtyq2, Rubber ducky222, Ruchi tspl, Rudjek, Rudolph.cm, Ruhrjung, Rumkles, Run26.2, Ruy Lopez, Rwalker, Rwenonah, Rwolfe85, RxS, RyGuy17, Ryan032, RyanGerbil10, RyanParis, Ryanbowden, Ryanjunk, Ryanluck, Ryanmcdaniel, Ryland510, Ryulong, Ryz05, RickK, S.U.H.E.L., S2006causa, S4diStiC, S71elements, S91by, S@bre, SCEhardt, SDJ, SDY, SEWilco, SFrank85, SG, SGBailey, SJMNY, SMC, SNIyer12, SPQRobin, SPUI, SS Division Wiking, SSZ, STARWARSGEEK, SWBJACMK, SWalkerTTU, SYSS Mouse, Sa.vakilian, Saashaw, Saathoff37, Saberswordsmen1, Sabrown100, Sabs1285, Sadads, Sadasd, Sade, Sadow300, Saforrest, Sagaciousuk, Sahrin, Saigon76nyc, Saksham, Sakutdjica, Salsb, Sam Coutu-Oughton, Sam Korn, Sam1993, Sam3199, SamEV, SamForestell, Samantha555, SameOldSameOld, Samir, Sammo, Sampo Torgo, Samroundface, Samuel Blanning, Samuel Sol, Samuelsen, Samwb123, Sanantonioguy, Sandersonjoe229, Sangjinhwa, Sango123, Sangre viento, Sannse, Sansvoix, Santiagobernstein, Sarah, Sardanaphalus, SarekOfVulcan, Saric, Saros136, Sasquatch, Sauerjos, Sauve.d, Savefrance, Savidan, Sayden, Saypotatoes8, ScAvenger, Scammerboy227, Scanlan, Scarab12, Scarian, Sceptre, Schadenfreude52, SchfiftyThree, Schmloof, Schnitzelman, Scholarly Breeze, Schoolsurfer, Schrödinger's cat, SchuminWeb, SchutteGod, Schutz, Schzmo, ScientificDude, Scientiom, Scientizzle, Scientus, Scientz, Scifiintel, Scifiradioguy, Scipius, Scirocco6, Sciurinæ, Scohoust, Scoobycentric, Scorchy, ScorpiOn, Scotchy500, Scott Burley, Scott Illini, ScottMainwaring, Scottmsg, ScriptBlue, Scrumy21, Scuppers1, Sdgjake, Sdhonda, Sdnegel, Seablade, Sean, Sean Whitton, Sean gorter, SeanO, Seaphoto, Search4Lancer, Searchme, Seb az86556, Sebbe, Secret (renamed), Secret Saturdays, Secretlondon, Seeraamaazu, Sekolov, SemDem, Semiquaver, Sennsa, Senorquigles, Separa, Sepeople, Seqqis, Seqsea, Seresin, Seric2, Serlin, Serminigo, Serpentinite, SerpentsTail, Sesel, Sesu Prime, Seth Ilys, Settl746, Sevillano1, Sexxie grrl, Sexy sally, Señorplayero, Sfacets, Sfahey, Sfan00 IMG, Sfmontyo, Sgguitarist, Sgodth, Shade11sayshello, ShadowGuardian, Shadowboy667, Shadowjams, Shadowy Crafter, Shaggy2Dope, ShakataGaNai, Shaken Earth, Shalom93, Shame On You, Shanedidona, Shanel, Shaneomac8605, Shanes, Shanesterman, Shannon7mcnamara, Shared Thought, Sharkticon Guy, ShaunES, Shawn m r, Shearonink, Shech736, SheikYerBooty, Shelleycakes, Sherzo, Shii, Shim jb, Shimonnyman, Shindleria235, ShipperS7, Shizhao, Shmew, Shoaler, Shoeofdeath, Shohei Okada, Sholtar, Shootappliances, Shoreranger, Shorne, ShortShadow, Shot007, Shpilo, Shredder2012, Shreshth91, Shroudan, Shunt110, Shushruth, Sicamous, Sicilianshotgun, Sietse Snel, Sig0, SigPig, Sigma 7, Sigmaparadox, Signsolid, Silence, SilentOpen, Silver Sonic Shadow, Silverchemist, Simhedges, Simishag, SimonP, Simulcra, Sinmiedoanada, Sir Nicholas de Mimsy-Porpington, Sir smorgasbord, Sir48, SirGrant, Sirkevinalot, Siroxo, SivaKumar, Sj, Sjakkalle, Sjö, Sk.Sajjad, Skaakt, Skapur, Skarebo, Skate4me, Skateboarderf2, Skeeter08865, Skentelbery, SkepticBanner, SkerHawx, Skier Dude, Skizzik, Skomorokh, Skooing, Skullfiend, Sky Divine, Skydeepblue, Skyking360, Skylinegtrgsl, Skyoon, Skyring, Sladuuch, Slakr, Slashcoconut, Slatersteven, Slconnor uk, Slcut, Sleptrip, Sligocki, Slivertransplant, Sloman, Slowking Man, Slrubenstein, Slugmaster, Slurpy121, Sm8900, Smaug123, Smcmanus, Smilemean, Smiling Kevin, Smiloid, Smith03, Smith2006, Smoke., SmokeyVillain, Smokeythecat, Smr16, Smug Irony, Snailwalker, Snoozer347, SnowFire, Snowmanradio, Snowolf, Snoyes, SoMuchTime, Soakologist, Soccerisawesome, Sodzei, Sole Flounder, Solipsist, Solitude, Somatosis, Some jerk on the Internet, Somedifferentstuff, Someone else, Somercet, Somethingoranother, Sonicfire3000, Sonjaaa, Sonyray, Soooclose, Soosed, Sopher99, Sophie2895, Sopranosmob781, Soulpatch, Soulstice, Soumyasch, Soundslikealotofhooplah, South Bay, SouthernComfort, Southsydney, Soxwon, Sp33dyphil, SpNeo, Space25689, SpaceFlight89, Spaceboy492, Spangineer, Spark98, Sparrowman980, Spartan078, Spartanloser, Spebudmak, Speedboy Salesman, Speedy06car, Spellmaster, Spencer195, Spettro9, SpiderMMB, Spiesr, Spiff, Spike105, Spikethehawk, Spinoff, Spiralhighway, Spitfire19, Spitzl, Splash, Splat, Spliffy, Sportsjam3, Spudtater, Spundun, SpuriousQ, Squash Racket, Squideshi, Squoups, Sri 1988, Srikantkedia01, Srishti1984, Srithikdatta, Srs, Srujanjoshi, Srushe, Ss2343, Ssilvers, Ssolbergj, Sssssss340, Sswonk, St00pidmonkey72, Stabby Joe, Stallions2010, Stan Shebs, StanZegel, Staplegunther, StarScout0304, Stars999, Starwrath, Stasven, Statalyzer, StateCollege101, Static11, Statsagogo, Staxringold, Stdevin2012, Steel, Stemonitis, Stenen Bijl, Stephen Gilbert, Stephenb, Stepshep, Stevage, Steve the Beach, Steve03Mills, Stevec240, Steven J. Anderson, Steven407653, Stevertigo, Stevietheman, Stevo1000, Stickee, Storasyster, Stormie, Stormscape, Strangepalefighter, Stratman07, Streryan, Strideranne, StringRay, Stroganoff, StrongProudGunny, Strongbad1982, Strovolos01, Stubblyhead, Student7, Stuman12, Stumink, Stupidstupidstupid, Stwalkerster, Subseven26, Substancep, Subsurd, Success123$, Sudeepneelakantan, SudoGhost, Suffusion of Yellow, Sufi dog, Suisui, Sumahoy, Sumanch, Sumone10154, Sunil060902, Sunquanliangxiuhao, Sunray, Sunshineofyourlove12, Supaman89, SuperDude115, SuperSmashBros.Brawl777, Supercoop, Supergeo, Supergodzilla2090, Superm401, Supersexyspacemonkey, Supertask, Superzoulou, SupremeRuler, Surcouf, Surgemg, Surlyduff50, Suruena, Suryoye, Susurrus, Suto, Sverdrup, Swainstonation, Swamilive, Swarm, Swatjester, Swedish fusilier, SwedishConqueror, SwedishZeta, Sweetback, Swellman, Swid, Swimm212, Swissbrazilian, SwnyMac, Swollib, Swordhunter0690, Syndicate, Synthe, Syrthiss, Sysrpl, Systemshocked, Syvanen, Szajci, Szakharin, Szyslak, T-rex, T1980, T800m101, TBrandley, TEG24601, TEMcGee, TFNorman, TFOWR, THEemu, THEunique, TJ Spyke, TJDay, TJive, TMC, TOMMASO, TOO, TORR, TR5000, TSO1D, TShilo12, TTBMA, TUF-KAT, TV1981HS, TXLonghorn, Ta bu shi da yu, TaRiQ 8d8, Tabletop, Taborgate, Tachyon01, Taco325i, TacticalMouse, Tad.k.22, Tafinucane, Tahir mq, Tai112, Taifarious1, Taikodrumfun, TakuyaMurata, Tamara Ustinova, Tangotango, Tanishh, Tannin, Tapir Terrific, Tapir2001, Tapirs65, Tardis, Targeman, Target Jackson, Tarheelz123, Tariqabjotu, Tarret, Tasc, TastyCakes, Tavatar, Tawker, Tayyabbaig, Tazchook, Tbhotch, Tbone55, Tbonnie, Tbsea, Tburgtbone06, Tdadamemd, TeaDrinker, Team6and7, Teammm, Techdawg667, Techguy118, Technical 13, Technocratic, Teddey, Tedickey, TehPlaneFreak, Tehmadhatter, Tekto9, TeleComNasSprVen, Teletoonmaniac, Telex, Tellsgeschoss, Telsa, Telstar2626436, TempDog123, Tempodivalse, TenOfAllTrades, Tenunda, Terence, Terfili, Terlynn4, Terra Novus, Terrasidius, Teryx, Testes rodriguz, TexasAndroid, Texaswebscout, Texinian, Texink, Texture, Tezero, Tf157, Tfine80, Tgannon, Tgeairn, Tha Nad 2007, Thadkobylarz, Thane, Thargor Orlando, Tharnton345, That-Vela-Fella, Thatdog, Thatotherperson, Thatpacokid, The Catholic Knight, The Cunctator, The Disco King, The Epopt, The Evil IP address, The Evil Spartan, The Four Deuces, The Great Liberator, The Halo, The Hitch22, The Land, The Lizard Wizard, The Man in Question, The Minister of War, The Number Man, The Ogre, The Placebo Effect, The Punk, The Revert States of America, The Tetrast, The Tom, The Transhumanist, The Universe Is Cool, The Utahraptor, The Wingman Joker, The ed17, The last Wikibender, The lost library, The machine512, The monkeyhate, The slave to reality, The ultimator, The wub, The-Holy-Hand-Grenade-Of-Antioch, The1edit, TheArmadillo, TheDoctor10, TheFarix, TheFuzzyFive, TheGWO, TheGrappler, TheHoosierState89, TheKMan, TheLongTone, TheMadBaron, TheNewPhobia, TheObtuseAngleOfDoom, ThePenciler, TheRanger, TheRomanLegion, TheThomas, TheTimesAreAChanging, TheVirginiaHistorian, TheWama, Theboondocksaint, Thebrother, Thechinese8, Thecrazydonvingigante, Theda, Thedec, Thedragonking, Thegargantua, Thegreatgrabber, Thegryseone, Thejason, Thelacker, Thelema418, Themane2, Themanwithoutapast, Themanwiththefish, Thematt523, Thenderson1, Thepebble3, Theredhouse7, Therequiembellishere, Theresa knott, Theriusthechronicler, Thermionic1, Thernlund, Therock40756, Thesevenseas, Thesouthernhistorian45, Thethirdperson, Theurgist, Thewolfstar, TheyMadeMeDoIt, Thezirk, ThievingBeagles, Thingg, ThinkBlue, ThinkEnemies, Thisbites, Thiseye, Thomas Paine1776, Thomas S. Major, ThomasK, Thomasjamesfoster96, Thomasmallen, Thomasmatoushek, Threepwood89, Thricecube, Thryduulf, Thunder8, Thunderbird2, Thunderboltz, Thurifer, Tide rolls, TigerShark, Tigey, Tightguy400, TigranTheGreat, Tim Starling, Tim Thomason, TimBentley, TimShell, TimVickers, TimothyHorrigan, Timrem, Timrollpickering, Timwi, Tintero, TintininLisbon, Titch Tucker, Titoxd, Tiwonk, Tiësto, Tjl1128, Tjmayerinsf, Tjoe, Tjpark1111, TkerTimeSeeker, Tktktk, Tktru, Tkynerd, Tlogmer, Tma 88, Tmazhindu, Tobby72, Tobycat, Tobyk777, Tocino, Tom, Tom Reedy, Tom harrison, Tom-, Tom8866, TomStar81, Tomass288, Tombyt55, Tomdatom, Tomgibbens, Tominator93, Tomlillis, Tommybtennis, Tommybtennis2, Tommyjb, Tomwood0, Tomwsulcer, Tony Sidaway, Tony age, Tony1, TonySt, Tonye104292, Tonywiki, Toonlad9, TopAce, TopazSun, Topbanana, Toropop, Tos, Tosfan4ever, Totla, Toussaint, Tparkes, Tpbradbury, Tpetross, Traal, Tragicfame, Trainspotter, Trannyboi, Transcendence, Trasman, Trav230, Travelbird, Traversc, Travis Cleveland, TreXiaNos, Tree Biting Conspiracy, Tregoweth, Treisijs, Trevdna, Trevor MacInnis, TreyHarris, Tribulation725, Triggerhippie4, Trinexx56, Trip Johnson, Tripacer99, Trongphu, Trontonian, Trottsky, Trovatore, True 505, Truecoder777, Trutal, Truthanado, Truthy not fakie, Ttwaring, Tubular, Tuckerdaman, Tud123, Turel, Turgeis, Turian, Turkuun, Tushar858, Tuspm, Tutmosis, Tuupu, Tverbeek, Twin Bird, Twirk88, Twump, Tx17777, Tycoon24, Tyhlerb, Tykell, Tyler McHenry, Tyrol5, Tysalpha, U S Crusher, UBeR, UH Collegian, UbZaR, UberBicep, Ubiquinoid, Ucanlookitup, Ucucha, Uebergeek, Ufwuct, Ugen64, Uglymoog, Uikos, Ulises Heureaux, Ultimus, Ultron, Umedard, Una Smith, Uncle J, Uncle Milty, Uncle Scrooge, Undernage, Universe=atom, Unixxx, Unsc, Unsungzeros, Unyoyega, Uogl, Upper Lines, Uppie, UpstateNYer, Uris, Url5406, Ursutraide, Uruguayo, Us441, Usaf2222, Usaforrest, UsefulWikipedia, User F203, User2004, Usergreatpower, Usfirstgov, Utcursch, UtherSRG, Uuu87, Uv411, Uxejn, V6g3h7, VLADIMIR Skokan, Vacilando, Valdezmcc, Valentino, VampWillow, Van helsing, VanBuren, Vandaltroll, Vanibal2, Vanished User 3388458, Vanished user 19794758563875, Vanished user 5zariu3jisj0j4irj, Vanished user 8jq3ijalkdjhviewrie, Vanished user 90345uifj983j4toi234k, Vanished user ewfisn2348tui2f8n2fio2utjfeoi210r39jf, Vanished user ikijeirw34iuaeolaseriffic, Vanished user kiij3irj4tihns, Vanished895703, Vaquero100, Vargob, Varlaam, Vary, Vcb, VeLi33, Vectoor, Veeraane, Veertlte, Vega84, VegaDark, Vegaswikian, Veghead, Velho, Velvetron, Venkat ramkumar, Ventagro, Vera Cruz, Verdadero, Vergetorix, Verrai, Verwoerd, VeryVerily, Verybigfish86, Viajero, Vicarious, Vicki Rosenzweig, VictorD7, VideoMan, Vikedal, Vilallonga, Vilerage, Villasmilraven, Vin Kaleu, Vincent Ricci, Vinceouca, VinnyCee, Viogfernos, Violask81976, Violetriga, Virgile1991, Virgule82, Viridae, VitaleBaby, Vitoghf, Vitr0x, VividTara, Vladdraculdragon, Vladmirfish, VolatileChemical, Voltagedrop, Votekyle, Vrac, Vranak, Vrinan, Vsmith, Vulpesinculta51, Vuo, Vzbs34, W.D., W00kie, W123, W2ch00, WGee, WHEELER, WJBscribe, WLRoss, Wackojacko1138, Wackymacs, WadeSimMiser, Wafulz, Waggers, Wahuskies96, Waja, Wakavankhai, Walkiped, Waltersimons, Wambeter, Wangi, Wangry, Wangyunfeng, WarEagleTH, Warbo5, Warset, WartyX, Wassamatta, Wassup234, Wassupwestcoast, Wastful, Watch37264, WatsonE, Watti Renew, Wavelength, Wawat, Waxxup, Wayward, Wbush89, Weaponx 2.0, Weavale, Webdinger, Weeliljimmy, WegianWarrior, Weien, Weinerheadshaboom, Wen Hsing, Wenteng, Werdan7, Weregerbil, Werideatdusk33, Werieth, Wes!, WesleyDodds, Wester, Westnest, Wetodid, Weyes, Wfaulk, Whalleyj, Wharrel, Whatthree16, Wheathampstead Brko, Wheelboss, Where, Whiskey in the Jar, Whisky drinker, WhisperToMe, Whiteboywonder, Whitedwarf1.3, Whkoh, Who, Whodoesntlovemonkeys, Whoop whoop pull up, Wickb55, Wickethewok, WideArc, Widefox, Wiggles (usurped), Wiglaf, Wiki alf, Wiki evan, Wiki man55, Wiki1609, WikiDao, WikiFlier,

WikiLaurent, WikiWitch, Wikiacc, Wikibofh, Wikid77, Wikidemon, Wikih101, Wikih8lulz, Wikihistorian, Wikiklrsc, Wikipedian06, Wikipedian77, WikipedianMarlith, Wikipedianinthehouse, Wikipelli, Wikitigger, Wikiwert, Wikiwikifast, Wikizach, WildlifeAnalysis, Will Beback, Will Colwell, WillMak050389, WillOakland, WilliamThweatt, Williamscold, Willingham, Willtron, Wils4581, Wilson44691, WilyD, Wimt, Winkelthorp, WinterSpw, Wipe, Wisco, Wisconsinhockey, Wisdom89, Wise, Witchzilla, With goodness in mind, Withouthavingseen, Wjfox2005, Wkcp, Wkdewey, Wknight94, Wkstar, Wog274643, Wolf ODonnell, Wolfman, Wonderboy2, WoodElf, WoodenTaco, Woodstone, Woohookitty, Woozle, Work permit, Working for Him, World, World-os.com, WorldTraveller101, Worldtimezone, Worrydream, Wpiki, Wrad, Wtmitchell, Wv235, Ww2censor, Wxlfsr, Wōdenhelm, XGustaX, XJaM, XLerate, XU-engineer, Xanzzibar, Xaosflux, XavierGreen, Xblahx, Xen0blue, Xenus, Xezbeth, Xiggelee, Xkcd, Xkcer Man, Xlegiofalco, Xloveangelmusicbabyx, Xoder, Xolom, Xooxgmoneyxoox, Xrblsnggt, Xrobx921, Xtreambar, Xwoodandwater, XxTrillvillexX9, Xyzzyva, Y2kcrazyjoker4, YAKNOWDOUGAL!, YORKABE, Yahoowill54321, Yahyawahbeh, Yair rand, Yakuzai, Yama Thi Khuu, Yamamoto Ichiro, Yamla, Yamummasnotbad, Yandman, Yangon, Yaninass2, YankeeDoodle14, Yankees4919, Yanks724, Yanksox, Yansa, Yardalhirji, Yashgaroth, Yasis, Yath, Yeehyunc, Yelgrun, YellowMonkey, Yelyos, YianniCart, Yitzhak1995, Ylai, Ymblanter, Yoganate79, Yohann4, Yojizu, Yongbyong38, Youhavebeentricked, Youndbuckerz, YourLawyer, Youssefsan, Yozer1, Ytmnd6, YuMaNuMa, Yubersoz, Yuckfoo, Yuje, Yupik, Yuri Gouveia Ribeiro, Yuusha, Z.E.R.O., ZICO, ZStoler, Zabararmon, Zach Vega, ZachPG, Zachary8222, Zachlipton, Zachorious, Zafiroblue05, Zahid2005, Zaksouthward, Zaldax, Zanimum, Zaphraud, Zarex, Zargontapel, Zarxos, Zauggb, Zayani, Zazaban, Zbxgscqf, Zealotgi, Zeamays, Zebe, Zeldaluver100, Zenohockey, Zephyr911, Zeppelin4life, Zeppelinrules, Zetterberg40, Zh, Zia demion, Zibzibzib, Zigger, Zloyvolsheb, Zmaher, Znarfaggle, Znode, Znorkel, Zoe, Zoicon5, Zoidy-Poo, Zondor, Zoney, Zoostar, Zotel, Zouavman Le Zouave, Zpowers, Zscout370, Zsinj, Zukariaa, Zvar, Zzuuzz, Zzyzx11, Zzzzzzus, Ævar Arnfjörð Bjarmason, Éponyme, АбуУбайда, ТимофейЛееСуда, مسلاب, सो ऽहम, , 5719 anonymous edits

Kentucky *Source*: http://en.wikipedia.org/w/index.php?title=Kentucky *Contributors*: 10.26, 16@r, 170.20.175.xxx, 1exec1, 1oddbins1, 2602:306:3BEF:3640:2055:5979:576A:D8F0, 2602:306:CFC9:2260:6233:4BFF:FE14:409C, 2602:30B:82B1:5339:38EC:9EBF:98C8:7336, 2help, 4twenty42o, 55david, 88AdolfLover88, 90, A2Kafir, A3r0, ARMYxTyler, AVand, Abductive, Acdixon, Addison0426, Addshore, Adrian.benko, After Midnight, Agrant131211, Agriculture, Ahassan05, Ahoerstemeier, Ahurt22, Airplaneman, Aitias, Aivazovsky, Ajenchel, Al Silonov, Alan Canon, Alanscottwalker, Alansohn, Alexa411, Alexf, AlexiusHoratius, Alexwcovington, Algebra101, Allstarecho, AllyUnion, Alpha 40475, AlphaEta, Amazingaro, Amazonien, American2, Amikeco, AnalGiraffe, Analogue Kid, Anaraug, Anarchy2190, Anaxial, Andrewman327, Andy M. Wang, Andy Marchbanks, Andy120290, Andyjsmith, Animum, Anna Lincoln, Anonymous555498, Antandrus, Anthony, Apusc, Arakunem, Arctic Kangaroo, Asa Miller, Assed206, Astuishin, Atlantia, AussieLegend, Aztom2, B, B-watchmework, BD2412, BMetts, BSveen, BabuBhatt, Bailey3475, Barek, Baronlewis, BarrelProof, Bassbonerocks, Beard rrp, Beland, Belligero, Ben Ben, BenBaker, Bender235, Benhealy, Benjh40, Big iron, Billyjoekoepsel, Bit72101, Bje2089, Bkell, Bkessler23, Bkonrad, Blaise170, Blehfu, Bloo Aardvark, BlueAg09, BlueAzure, BlueMoonlet, Bluedenim, Bmpeers, Bms4880, Bmv 1978, Bobblewik, Bobbo, Bobbyllama, Boblarazzi, Boblarazzi2, Bobo192, Bolivian Unicyclist, Bomac, Bongwarrior, Boom 2010, Borderer, BoycottKY, Bradlydollar, BrainMarble, Brando03, Brandon, Brbigam, Brian Schlosser42, Brighterorange, Brutannica, Bryan Derksen, BryanG, Bryanwillett, Bsmeathers, Buaidh, Bulba2036, Bullshark44, BuzyBody, C.Fred, CJLL Wright, CJR127, CQ, CSWarren, Cadiomals, Caiaffa, CalicoCatLover, Caliga10, Camarcri000, CambridgeBayWeather, Campaigner444, CanisRufus, Canterbury Tail, Caponer, Capricorn42, Casper2k3, Cburnett, Cchow2, Censusdata, CesarB, Cfrydj, Chanheigeorge, Charles Wilton, Charlescaylor, CharlieEchoTango, CharlotteWebb, ChocolateSaltyBalls, Choichy5, ChongDae, Chris the speller, Chris24, ChrisGualtieri, ChrisRuvolo, Chrislk02, Christian145678, Christian75, Chun-hian, Cinik, Circeus, Civil Engineer III, Ckatz, ClairSamoht, Clearlyme, Cliffb, Cmadler, Coasterlover1994, Codydrew, Coemgenus, Coine in the pocket, Cometstyles, CommonsDelinker, Compfreak7, Conankudo51, Confiteordeo, Conversion script, Cool Blue, Cornishgame, Cougar11, Courcelles, CrazyC83, Crobertson, Crunchy Numbers, Cuchullain, Cwalsh7997, Cwiki, Cyanidethistles, CylonCAG, Cyndi63, D6, D9526328443789, DARTH SIDIOUS 2, DCEdwards1966, DDear99, DE, DECOR8Rgirl, DIREKTOR, DJBullfish, DVD R W, DVdm, Da Packman, Dabomb87, Daddavout, Dale Arnett, Dalliance, Damaster9106, Daniel Case, Danny, DarkFalls, Darkness Productions, Darksupertailsisborn, Dave, Daven200520, David.Gaya, DavidHarden, Dawn Bard, Dawnseeker2000, Day44581, Dblevins2, Dcolemire, Dcornwall, Dcsohl, Debresser, Deconstructhis, Deflective, Delldot, Deltabeignet, Derek.cashman, Deville, Dick Kimball, Dietcokementos, Disavian, Discospinster, Dkdantastic, Dlh89, Don de la Muncha, Doncram, Doniago, Doom2000, Doulos Christos, Douschonmandick, Download, Dp462090, Dr Nyan, Dr. Blofeld, Dralwik, Drano, Drclay88, Drmies, Drumex, Dtox11, Dubyavee, Duckwing, Dweller, Dwo, Dysprosia, ERcheck, ESkog, Eastcote, Eastfrisian, Eco84, Ed g2s, EdBever, Edison, Editmewiki, EditorMan ha ha ha, Edivorce, Edward, Ejc891, Ejrubio, Ekren, ElKevbo, Eldude611, Ellywa, Elm-39, Emhoo, Emmette Hernandez Coleman, Enauspeaker, EncMstr, Encephalon, EncycloPetey, Engineer Bob, Epbr123, Epicgenius, Epolk, ErgoSum88, Eric, Ericnease, Erik Kennedy, EurekaLott, Evans1982, Everyking, Evice, Evrim98, Ewen, Excirial, Explicit, FPAtl, Falkirks, Faradayplank, Fartman298, Favonian, Figgie123, Finalius, Fishing, Fizzleshizzle, FlatFork, Flatterworld, Fledgist, FloNight, Floridahistory, Fogherty V. Tatin, Folic Acid, Footballfan190, Footwarrior, Fourthords, Fradd, Francs2000, FreplySpang, Frietjes, Funandtrvl, Funnyfarmofdoom, Funnyhat, Furado, G.Freeman123, GM11, Gaius Cornelius, Gatita09, Gator87, GeorgHH, George Ho, GeorgeLouis, Gerry D, Gfoley4, Gilliam, Gimboid13, Gimmetrow, Giraffedata, Glacialfox, Glass Sword, GlassCobra, Gmaxwell, Gmhetrick, Gogo Dodo, GoingBatty, Gold heart, Good Olfactory, GraemeMcRae, Grafen, Graham87, Grayson50000, Great50, GregAsche, GregU, Greswik, Gsmgm, Gtbob12, Gurchzilla, Gurt Posh, Gymnast, Gökhan, HJ Mitchell, HMSSolent, Haakon, Had22, Hafrolocks, HalfShadow, Hallel, HamburgerRadio, Hamtechperson, Happyfatman021, Happysailor, Haymaker, Heegoop, Hellogj, Hellomynameisjoswai, Hellosandimas, Hendo1270, Hertz1888, Heuertag, HexaChord, HiB2Bornot2B, Hippymac19, Hmains, Hobartimus, Hoke001, Hottentot, Hu12, Huntington, I dream of horses, IFCAR, IRP, Ianwhite21, Icedman42, Immunize, Imroy, Incornsyucopia, Instinct, Invisbob, Iridescent, IronChris, Ironholds, It's-is-not-a-genitive, Itascapark, Itfc+canes=me, Itinerant1, Ivirivi00, Ixfd64, Izno, J.Steinbock, J.delanoy, J36miles, J654567, JCO312, JHunterJ, JMRG, JNRVM, JS Nelson, JW1805, Ja 62, Jaardon, Jab843, Jack Cox, Jack Merridew, Jacob.jose, Jajhill, Jakedabeast, Jaknouse, James Byrne69, James Nicol, JamesBWatson, Jamison1234, January, Jasonestevan, Jasper Deng, Jatkins, Jauhienij, Jaw959, Jaypur, Jaysing103, Jaywubba1887, Jbamb, Jbfwildcat, Jcembree, Jcravens42, Jduckeck, JdwNYC, Jeffrey O. Gustafson, Jengod, Jfioeawfjdls453, Jhendin, Jiang, Jill Orly, JimIrwin, JimVC3, JimWae, Jimp, JinJian, Jm0371, Jmeisenhelder, John K, John Reaves, John254, JohnInDC, JohnOwens, Jojhutton, Joncnunn, Jopjopjop123, Jose77, Joseph Dwayne, JoshEdgar, Joyous!, Jsayre64, Jschnur, Jspivak86, Jsum, Jt 200075, Juliancolton, Jultemplet, Junglecat, Juniebjone, Jusdafax, Just Say No, Juzeris, KGasso, Kahuna207, Kahuroa, Kaos456, Karam.Anthony.K, Katalaveno, Katherine, Katieh5584, Kct124, Kd4nuh, Keegan, Keilana, Keith Edkins, KeithH, Kem0106, Ken Gallager, Ken g6, Kenaldinho10, Kentucky1234567890, Kentucky1333, Kentuckyliberation, Kevin B12, Kevin Myers, Kevindl85, Kguiver, Khatru2, Khazar2, King of Hearts, Kinneyboy90, Kinston eagle, Kintetsubuffalo, Kittysrfun, Kleinklein, Knowledge Seeker, KnowledgeOfSelf, Koavf, Kooldude28, Kozuch, Kpjas, Kralahome, Kralizec!, KramerNL, Krenair, Kronos o, Kumioko (renamed), KurtFaulkner, Kvasicek, Kwamikagami, Kyace, Kylepotter, KyraVixen, Kyshooter, L Kensington, L.cash.m, La goutte de pluie, LaOtto, Lamontacranston, Landroo, Lankiveil, Law, LeaveSleaves, Lebanonman19, Ledenierhomme, Lemonflash, Leo1410, LeoDV, Les Invisibles, Letcher Mountaineer, Leuko, Levelistchampion, Levi93, Levineps, Lightmouse, Lilac Soul, Ling.Nut, Little green rosetta, LizardJr8, Lizmichael, Logan, LonghornsU10, Lord Emsworth, Lord Voldemort, Louisvillian, Lreid0975, Lucidwave, Lucky Mitch, Lugia2453, Luna Santin, M C Y 1008, MER-C, MJCdetroit, MONGO, MPF, MPerel, Macarenses, Mad Lothairian, Maddie!, Madisonhenry46, Maelnuneb, Malathos, Malcolm Farmer, Mamyles, Mandall 32, Mandarax, Marek69, Marketinggal, Marmote, Martin451, Marysunshine, Massjit, Mastrchf91, Matankg, Matijap, MattTM, MauriceHall, Mav, Maxtremus, McCorrection, McSly, Mdmorgan, Meaghan, Meelar, Meepmeepmany, Melsaran, Meltonkt, Metalhead94, Mhbourne, Mhudson2, Miamimusicpro, Michael Devore, Michael Simpson, Midnight Green, Mike s, MikeWilson, Mikeo, Mikevegas40, Mindcry, Mini LJ 69, Minimac's Clone, Mintleaf, Miskwito, Mister Alcohol, Mitchhelper, Modulatum, Moe Epsilon, Monocletophat123, Moreau36, Moverton, Moxy, MrBleu, MrFish, Mrt3366, Mrwiki355, Mu Cow, Mufka, Mulad, Murraytest, Musicgeeksquad, Mwanner, Mxn, Myfatchildisscary, Myweenieiseenormus, N Vale, N734LQ, NCDane, NE2, NEICenergy, NJA, Nahoyman78, Nakon, Nascar1996, Nasnema, NawlinWiki, Neelix, NellieBly, Netparrot, Neutrality, News Historian, Nick Comber, Nick Number, Nihiltres, Ninly, Nirvana23m, Nivek1385, Nivix, Nnemo, Nneonneo, No such user, Nomatterwhereigo, Noraft, North Shoreman, Nskillen, Ntsimp, Num1scot, Nuno Tavares, OOODDD, OPMaster, Oda Mari, Ohnoitsjamie, OklahomaSwag, OlEnglish, Old Abe, OmeiKida, Ommnomnomgulp, Omnicog, Omygawsh, Ondewelle, Oroso, Orphan Wiki, Ost316, OutrunA1C, OverMyHead, OverlordQ, OwenX, Oxymoron83, Ozi, PBP, PCHS-NJROTC, PETEYTHEREF, PFHLai, PL290, PMDrive1061, Pan Dan, Pankkake, Pantallica megadeth, Parkwells, Pass a Method, Passionless, Patrick, Patrickneil, Patricknoddy, Patronik, Pccromeo, Pearle, Peets, Pejman47, Pennst110, Penthrift, Perrytwinkle, Persian Poet Gal, Perthsider, Petrb, Petri Krohn, Petropoxy (Lithoderm Proxy), Pfly, Pgk, Phase Theory, Philip Trueman, Phox6469, Phung Wilson, Piano non troppo, Pigman, Pinethicket, PlanetBoyer, Plasticup, Pluma, Pmac21, Poccil, Poepla, Pollinator, Polymerbringer, Possum, Postdlf, Prodego, Proficient, Pshent, Psthi, Pvmoutside, Qqqqqq, Quadell, Quantpole, Queson, QuixoticLife, Qwyrxian, R'n'B, R-HIT, R.T.Gellar, RHaworth, RL0919, Racerx11, Ragib, Railer 103, RainbowOfLight, Rakista, Ramrod1001, Randomfrenchie, Raprat0, Rarefly45, Ratsrule, Rattlesnakeryder, Ravedave, Raymiey, Rayne GS, Razorflame, Rd232, Rdikeman, Realm of Shadows, Reaper Eternal, Recognizance, Red Director, Red dwarf, RedCoat1510, Redsully, Reesh, Regoarrarr, Rentman, Retired username, Revas, RexNL, Reywas92, Rhatsa26X, Rhopkins8, Rich Farmbrough, Richard Keatinge, RichardF, Rick Block, RickK, Rividian, Rjensen, Rjwilmsi, Rlquall, Rmatsu888, Robert LeBlanc, Robomanx, RobyWayne, Rocketrod1960, Rod Foster, Roger231, Roger2518, Rogerd, Roleplayer, Romanm, Romannumeral, Romarin, Root Beers, Rossdegenstein, RoxyTheWolf, Royar, Rreagan007, Rtshawnee, Ruhrfisch, Rumping, RussBlau, Rustyjj70, Rwalker, Ryan032, S.ross5, S91by, SMC, Sahara4u, Sam Dempsey, Sam Korn, Samian, SamuelFreli, Santeboo12, Saopaulo1, Sarmatian07, Saturday, SatyrTN, Sayeth, Scanlan, Sceptre, SchfiftyThree, SchnitzelMannGreek, Schnoatbrax, SchuminWeb, Scjessey, Scootey, ScottSteiner, ScottyBerg, Sct112569, Scythian1, Seaphoto, Seb az86556, Seicer, Sessenem Kram, Sfmontyo, Sgr2000, Shadowjams, Shanes, Sharkface217, Sherf77, Shirik, Shlomke, Sierra, Simeon24601, Sinn, Sirrom nodnarb yerffej, Sjakkalle, Sk8er01, Skamecrazy123, Skankadoodles, SkerHawx, Skizzik, Skyezx, Slazenger, Smartie2thaMaxXx, Smashville, Smith03, Smokytop, Smylezsh, Snowolf, Sol!66, Solipsist, Some jerk on the Internet, SonPraises, Sonpraises, Sophus Bie, Sorabezhad, Sosasos2001, Sosthenes12, Southernky, Southronite, Sparsefarce, Spasemunki, Spydrlink, Squigglybee, Sreejithk2000, Stallions2010, Stan Shebs, Startstop123, SteelTownMoFo, Steelers130796, Stelio1234567890, Stepshep, Steve03Mills, Steven Gazeas, Stevewonder2, Stevietheman, Stevotower, StickerMug, Storkynoob, StuffOfInterest, Stuhacking, Swollib, Syra987, TBrandley, TKD, TUF-KAT, TacoMonkey1, Tamer of hope, Tannergantt12345, Tbhotch, Tdadkins1, Teammm, Teapot37, Techman224, Technopat, Tedickey, Teh tennisman, Template namespace initialisation script, TenTech, Tewy, The Epopt, The Punk, The Thing That Should Not Be, The Transhumanist, TheFox, TheHoosierState89, TheNewPhobia, TheYoungThinker, Thedesolatewest, Thehelpfulone, Thejackyl, Theredhouse7, Thesouthernhistorian45, ThievingBeagles, Thingg, Thunderectum, Thunderpooch, Tide rolls, Tidying Up, Timneu22, Tjparsons, Tmansv, Tobby72, Todd Vierling, Tom Allen, Tom mayfair, Tom.k, TomStike, Tomas417, Tomatosoup97, Tommy2010, Tony1, Totaldismair, Tow, Tpbradbury, Tracer9999, Transcendence, Treisijs, Turtle ball 2, Twaz, TwinCityIL, Twitmarinemonk, Ucucha, Ugur Basak, Uirauna, Uklf, Ulric1313, Ultra X987, Underwoodm, Useight, Utcursch, Uyvsdi, Vanished user ewfisn2348tui2f8n2fio2utjfeoi210r39jf, Vbofficial, Velvetron, Vendettax, Versageek, Victor, Vina, Viriditas, Vranak, Vsmith, Vzbs34, WHeimbigner, WJBscribe, WOSlinker, WPANI, WQUlrich, WadeSimMiser, WalkenIkayne, Walrusfighter, Walton One, Wangi, Wapcaplet, Warbuff 4, Wars, Wayne Slam, We are getting pist now, Weatherman78, Webclient101, Welshbear61, Weregerbil, Wghornsby, WhisperToMe, Whocares81, Whyarentthereanyusernames, Widr, WikHead, Wiki13, Wikicrasher, Wikieditor06, Wikiklrsc, Wikipedianinthehouse, Wikipelli, Wikipeterproject, Wildcat156, Will Beback, William Avery, Willking1979, Wknight94, Wmahan, Woohookitty, Wtmitchell, WxGopher, Wywin, X96lee15, Xaosflux, Xazbq, Xdenizen, XisTheBest, Xyzzyva, Yaf, Yamamoto Ichiro, Yandman, Yintan, Yloop, Youngamerican, YourNight, Yuckfoo, Zaui, Zntrip, Zoe, Zoicon5, ZooFari, Zzuuzz, Zzyzx11, Δ, 2046 anonymous edits

2000_United_States_Census *Source*: http://en.wikipedia.org/w/index.php?title=2000_United_States_Census *Contributors*: Aaron McDaid, Agtaz, Alansohn, Alexander Domanda, All Is One, Altenmann, Amaury, Andrea105, Arthur Rubin, Aude, Avenue, Bangaboy, Barrylb, Bart Versieck, Bender235, Betacommand, Big Bird, Binksternet, BlackMamba24601, Blanchardb, BlastOButter42, Bluelion, Bobblewik, BrendelSignature, Brockert, Bryan Derksen, Bubbles dude, Buster7, Buzzerzen, CJLL Wright, CP\M, Caesura, CanadianCaesar, Captain panda, Centrx,

Christopher Sundita, Circeus, Closeapple, Commodevoncommode, Compfreak7, CoolGuy, Courcelles, Cptmurdok, Cybercobra, Dale Arnett, Danthemankhan, Dark Tichondrias, Darkfight, Davidcannon, Davodd, Daysleeper47, Dedeededee23, Delldot, Democracy to information, Dman69420, Drizzle3d, Editpop, Enviroboy, Epbr123, Erik9, Erudy, EurekaLott, Everyguy, Facemask645, Farras Octara, Foofighter20x, Fpdeo45, Fraggle81, Freeman45Fighter, Furado, G1076, Gene Nygaard, Gilderien, Gnj, GoldRingChip, GregAsche, Ground, Guoguo12, Hadal, Heryu, Hmrox, Hornandsoccer, HotWheels53, Izanbardprince, J. Spencer, J.delanoy, JHMM13, JPG-GR, JWB, Jamesofur, Jason Palpatine, JayJay, Jeepday, JeffreyN, Jengod, Jerel411, Jimbo Wales, Jimothytrotter, Joncnunn, Jorge Stolfi, Jrob123101, Karenjc, Keith D, KenFehling, Kevthegreat55, KnightRider, Koosle, Lensovet, Leuko, Lgood29, Logan, LoverOfTheRussianQueen, Mack2, Mamyles, Mangoe, Manyspots, Mike Rosoft, Missmarple, Moonriddengirl, Mxn, Mygerardromance, Navy Blue, NickCT, Niteowlneils, O18, OwenBlacker, Pascal.Tesson, Pax:Vobiscum, PeriwinkleEnglish, Peruvianllama, Peteforsyth, Petri Krohn, Philip Trueman, Pigman, PredatorHunt791, Psiphiorg, Quazgaa, Radon210, Ran, Reinyday, RenamedUser01302013, Reywas92, Rgoodermote, Rich Farmbrough, Rjwilmsi, Rmhermen, Rockhead126, Rrius, Rugadh, Saehry, Safety Cap, SamEV, Sir Lothar, Slon02, Smalljim, Smittyredsox, SoCalSuperEagle, Stepshep, SteveSims, Sycomonkey, Tanthalas39, Tantus100, That-Vela-Fella, The Thing That Should Not Be, The ed17, TheFeds, TheGeneralUser, Thingg, Tide rolls, Towerdefence, Tpbradbury, Tribefan43725, Truthanado, Twinsday, UpstateNYer, User At Work, User5437, VengeancePrime, Vipinhari, Whitd48, Yamamoto Ichiro, Zbase4, Zigger, ZooFari, Zzuuzz, 239 anonymous edits

United_States_Census_Bureau *Source*: http://en.wikipedia.org/w/index.php?title=United_States_Census_Bureau *Contributors*: 007bond, Abductive, Acsenray, Aephyx, Ahoerstemeier, Al Lemos, Alan d, Alex1011, Alexandria, AndrusTandrus, Anna Frodesiak, AnonMoos, Antandrus, Armonica, Astrocommander, Asturianu, Atlien, Aude, AutoGeek, Babij, Beland, Bender235, Big Adamsky, Bikepunk, Bill.lay, Biruitorul, Bkonrad, Blargh29, Bonder00700, BradVanDam, Brandon5485, Brookie, Buaidh, Buster7, CMG, CambridgeBayWeather, Can't sleep, clown will eat me, Cancandance, CapitalR, Cartoonmoney, Chamal N, Chemboss, Chrism, Ckatz, Clayoquot, Clindberg, Closeapple, Cocircular, Coffee, Colonist, Compfreak7, Conversion script, CopperSquare, CsDix, Csyberblue, Curps, Cybercobra, Darkfight, Daven200520, David Kernow, Davilaa, Davodd, Den fjättrade ankan, Deodar, Der Träumer (DE), Dipankan001, Discospinster, Dmgaldi, Dough4872, Download, Dr.PeterRobertson, Dthomsen8, Duffman, EOBeav, ESkog, Eastlaw, Econr6, Eeaee, Elakhna, Epbr123, Epolk, EurekaLott, Evercat, Farquaadhnchmn, Favonian, Formulax, Framed0000, Frank101, Frazzydee, Furado, Fæ, GABaker, Gadickson, Gauth001, Geraldo Perez, Gidonb, Giftlite, Globalsolidarity, Gnominonwiki, GoldRingChip, Gpollock, Grafen, GregAsche, Ground Zero, Gugganij, Happenstancial, Harro5, Haruo, Hello71, Hephaestos, Hoppingalong, ImperfectlyInformed, Izzy007 1, JONJONAUG, Jengod, Jerryseinfeld, Jholman, Jim1138, Joao Xavier, Joseph Solis in Australia, Jusdafax, Jvcdude, Kbdank71, Ken Gallager, KenFehling, Klaam, Koavf, Kochamanita, Krellis, LOL, Lear's Fool, Leif, Leinad, Leszek Jańczuk, Levineps, LizardJr8, Lotje, Luna Santin, Magioladitis, Markjohnson87, Marshman, Martinwj53, Materialscientist, Mav, Meaghan, Mediran, Megaman en m, Michael Hardy, Microtony, Minesweeper, Musichman, Mxn, MyTwoCensus, Mysid, NickCT, NicoNet, Nightscream, NoSeptember, O18, Ohnoitsjamie, Olivier, One, Parkwells, Patrick, Pax:Vobiscum, Peak, Pearle, Pi, Pi zero, Pittimann DE, Pittimannl, Polmandc, Postdlf, Reach Out to the Truth, ReadQT, Rettetast, Revth, Ricav, Rich Farmbrough, RingtailedFox, Rklear, Robert Merkel, Rougher07, Rwwww, Sardanaphalus, Sarranduin, Scriberius, Semolo75, Seth Goldin, Shannon bohle, SimonP, Sk8tercolt18, So-nj, Spangineer, Srikeit, Srushe, Ssbohio, Stelmakia, Storm Rider, TJMcDJ, TOttenville8, Tanweer Morshed, Tedickey, Tenthfirst, Textorus, The Anonymouse, The Epopt, The Nut, Themfromspace, ThinkingTwice, Tide rolls, Tim!, Tisane, Tom, Tootallmaan, Touchmytralala, TygerDawn, UH Collegian, Vanka5, Ventur, Vuurcifer, Wars, Wayne Slam, West.andrew.g, WhisperToMe, Widr, WikiEditor5493, Willking1979, Yngvadottir, Yone, ZeWrestler, Zzuuzz, شرزر, 231 anonymous edits

Census *Source*: http://en.wikipedia.org/w/index.php?title=Census *Contributors*: 16@r, 2toise, 90 Auto, A. B., ASDFGH, Abhaga, Achim Jäger, Adam Bishop, Aewold, AgainErick, Ahmedrefat, Ahmet Huner, Ahuskay, Aitias, Alansohn, Alex.muller, Alexd321, Alexzr88, Allstarecho, Altenmann, Andres68, Andrewpmk, Andy Marchbanks, Andycjp, Angrytoast, Angusmclellan, AnonMoos, Antiuser, Arcadian, Argon233, Arkaig1, Arnab1996, Artaxiad, Ashley thomas80, Australia boy, Avala, Avenged Eightfold, Avoided, BD2412, Badanedwa, BanyanTree, Barefeetdude, Barkers PR, Bart Versieck, Bender235, Benlisquare, Berean Hunter, Berkunt, Bersus, Bffjil, Bfswebstat, Bhadani, Bigsean0300, Birnuson, Biruitorul, Bitbut, Blackknight12, Blehimbored:), Bloodshedder, Bobblewik, Bogdangiusca, Bongwarrior, Boomer1234567, Boreas74, Boxplot, Brandmeister (old), Brianjd, Brutalhovno, Bry9000, Bucketsofg, CRICKETLtd, CTF83!, Ca3tki, Calliopejen1, Calvintiger, CanOfWorms, Canadaonline, Carboli, CaribDigita, Carwil, Cdc, Celtmist, Cgingold, Cherkash, Chester Markel, Chirag, ChrisGualtieri, ChrisNickson, Chrislk02, Chrism, Christian List, Chrono1084, Chuq, Ckatz, Closedmouth, Cmapm, Cnees, Colchicum, CommonsDelinker, Conversion script, Coopkev2, Cordless Larry, Courcelles, Cpl Syx, Crazyquesadilla, Cribbswh, Cstreet, Cup o' Java, Curb Chain, Cyanoa Crylate, Cymru.lass, D. F. Schmidt, DBigXray, DS1953, Dale Arnett, Dan653, DanielCD, Datu Dong, David Thrale, David1217, Davidcannon, Dbachmann, Dcastor, Ddroar, Deeptrivia, Delirium, Denisarona, Dewet, Dgkuty, Die spam die, Dinakarr, Dirkbb, Discospinster, Doggitydogs, Dominictroc, Donald Albury, Donfbreed, Dongjing, Doric Loon, Doseiai2, Dramatic, Drozmight, Dub8lad1, Dwight666, EamonnPKeane, Earl Andrew, Edcolins, Edward, Eilthireach, Eleassar, EmanWilm, Emilios1310, Emperorbma, EqualRights, Eregli bob, Erianna, Eric Wester, Erkan Yilmaz, Estoy Aquí, EugeneZelenko, Evil Monkey, Exec8, Ezhiki, Facial, FakTNeviM, Farhan Zafar, Fatstraw, Favonian, FedLawyer, Fern Forest, Fg2, Flapdragon, Flying Saucer, FocalPoint, Footballover, Fratrep, Frazzle, Fredbauder, Freshbakedpie, Futurist110, Fz22, Fæ, GABaker, Gaius Cornelius, Galoubet, Garrettmca, Gazilion, GenericBob, Genev82, Geoffrie, Gidonb, Giraffedata, Gogo Dodo, Gointemm, Gookhatingvet, Gordon.lunar, GraemeL, Graham87, Granucci, Grick, GringoInChile, Ground Zero, Guest9999, Gun Powder Ma, H2g2bob, H@r@ld, HJ Mitchell, Hadal, Hagedis, Hairy Dude, Halibutt, Hank7hill, Hattrem, HenryLi, Hephaestos, Heyhey Hey, Hibernian, Hmains, Hohum, Homerjay, Hut 8.5, Hwy43, Hydrogen Iodide, IRP, IRWolfie-, Igoldste, Ihatepotsmokinghippies, Iiiiaaaa, IkonicDeath, Imoen, IndepIntel, Intgr, Iridescent, Isaac120, Isthmus, Itsmejudith, JForget, JHunterJ, JackofOz, Jagged 85, Jannismakris, Jauerback, JavierMC, Jboyle4eva, Jeandré du Toit, Jeff G., Jeraphine Gryphon, Jerry Ritcey, Jerzy, Jibril23, Jim Huggins, Jim no.6, Jimp, Jiri007, Jm1122k, Jnestorius, Joao Xavier, JoeSmack, John ellenberger, John of Reading, JohnGurney, Johnjay1745, JonHarder, Jonemerson, Jonesey67, JorgeGG, JorgeRodriguez, Joseph Solis in Australia, Joseph simeon, Josette, Julius Sahara, Julo, Jusdafax, Jusjih, Just plain Bill, Just4ikarus, KGasso, Kaihsu, Kat263, Kay556690, Kayau, Keith Edkins, Kembangraps, Khalil babli, Kiefer.Wolfowitz, Kigali1, Kim bas, King Lopez, KingKane, Kingpin13, KirrVlad, Kku, Koroner, Krikwen, Kristaga, Kross, Kyky108, L Kensington, Lacrimosus, Laogeodritt, Lasunncty, Lateg, Lauzonda, Lawlessurchin, Ldsviking, Legumeman, Leszek Jańczuk, Levineps, Li-sung, Licor, Lindar, LizardJr8, LlywelynII, Llywrch, Logicwiki, Ltool63, Luna Santin, Luntertun, LynxNukem, Mack2, MadGeographer, Magbatz, Malcolm Farmer, Mandarax, Mandsford, Manuel Anastácio, Marathone, Marek69, Mark Arsten, Martpol, Masoudfcr, Materialscientist, Matt Brennen, Matthead, Mattsains, McSly, Meiskam, Meister, Melcombe, Mentifisto, Mhym, Michael Devore, Michael Hardy, MichaelCarl, Midway, Mikkel, Milboz, Mild Bill Hiccup, MisfitToys, Misskaykay, MithrandirAgain, Mkanoni, Modulatum, Mr.Badlands, MrBell, MrFawwaz, Mrh30, Muppet, Mxn, MysticalCow, Mzajac, Mátyás, Nabob Nagusa, Nachoman-au, Nateofgreat, Nauseam, NawlinWiki, NeonMerlin, NerdyScienceDude, Nestled like spoons, Neurolysis, Newhoggy, Nfitz, Nhgenealogy, Ninsheart, Niteowlneils, Nmacu, Nn123645, Nomi887, Norbu19, Notheruser, Nponte, Nrlight, Ns.code, NuclearWarfare, Numbo3, Nunquam Dormio, Nyttend, O'Dea, Ohconfucius, Oleg Alexandrov, Oliver Pereira, Olyus, Omdfw, One, Onya4ever, Optimist on the run, Orodreth, PANONIAN, PMLF, PRODUCER, Panapp, PaoloValente, Parishan, PatrickFlaherty, Paul Norfolk Dumpling, Paul W, Paul-L, Pepith, Pgallert, Phantomsteve, Pharaoh of the Wizards, Philip Trueman, Philopp, Pigman, Pinethicket, Piotrus, Plastikspork, Pne, PoeticVerse, Polly Math, Poor Yorick, Priyansh verma, Prop3v56, Quadell, Quadparty, Quebec99, Queenmomcat, Quicksilvre, Qwertzy, Qwfp, RODERICKMOLASAR, RTG, Rami R, Rammy2cool, Razor2988, Reach Out to the Truth, Reaper Eternal, Red Slash, Reinyday, Remigiu, Ren Sydrick, RenamedUser01302013, Rentzepopoulos, Rhopkins8, Rich Farmbrough, Richard Arthur Norton (1958-), Rjwilmsi, Rmhermen, Rollon, Romanm, Ronhjones, Rui Gabriel Correia, Ryl411, Saddhiyama, Sahasrahla, Sandstein, Sannse, SatyrTN, Savidan, Scientus, ScottyBerg, Sdaeun, Seaphoto, Selket, SemperBlotto, Senator2029, Sesu Prime, Sfs90, Shadowjams, Shannon bohle, Shii, Shoessss, Siamax, Sift&Winnow, Simetrical, Sionk, Sivuman, Sjakkalle, Skizzik, SlimshadY, Smallman12q, Smarkflea, Smooth O, So-nj, Sombrero, Sortcoach, SpacemanSpork, Spearhead, Specs112, Spoladore, Srich32977, StAnselm, StaticGull, Stet, Sukarnobhumibol, Tarih, Tarun mor50002, Taztouz, Tempac, Terovian, The Rambling Man, The Thing That Should Not Be, TheRingess, Thunderbuck ram, Tide rolls, Tim1357, Timc, Timneu22, Timrollpickering, Tomcakes, Tomi, Tommy2010, Tomsega, Tony Corsini, Tpbradbury, Treisijs, Triona, Truthanado, Tualha, Twhyatt, Typing General, UberScienceNerd, Ultraboy8888, Uncle Dick, Unnikn, Urg writer, Utar, Valhalla, Vanished user 342562, Vicki Rosenzweig, VictorianMutant, Wavelength, Wayne Riddock, Wheatleya, WhiteWriter, Whodhellknew, Wienertakesall, Wiki-uk, Wiki1609, WikiPuppies, Wikidan829, Will Beback Auto, Wimt, Wolfvuk, Wtshymanski, Wysprgr2005, XNanakate, Xanthar, Ycu1101ycu, Yidisheryid, Yngvadottir, Ynhockey, Yom, Yoman82, YuMaNuMa, Yupik, Zaian, Zangar, Zigger, Zinc2005, Zlerman, Zollerriia, Zundark, Zxcvn80, Zzuuzz, نگورابانگوری، اشف عقیل, 847 anonymous edits

Population_density *Source*: http://en.wikipedia.org/w/index.php?title=Population_density *Contributors*: 1234r00t, 1297, 21655, 2mcm, 45Factoid44, 4halsop, A-giau, A. Parrot, A.Ou, Abhijitsathe, Achim Jäger, Addihockey10, Agentbla, Ahoerstemeier, Alansohn, AlexD, Alexjohnc3, Andre Engels, Andycjp, Anthonyz747, Aphaia, Apokryltaros, Applecharm, Aranel, Areology, Astuishin, Aude, Avicennasis, Aziz1005, B.d.mills, Barend, Barneca, Biederman, Blacklite, Blaircorrigan, Blanchardb, Bobblewik, Bogdangiusca, Br77rino, BrainyBabe, Brianlo747, Bricklayer, Brisvegas, BrokenSegue, Bsadowski1, Buchanan-Hermit, Bwilkins, CUSENZA Mario, Calabe1992, Candorwien, Catgut, Cdc, Cenarium, Chimpex, Ckatz, Cliff smith, Clovis Sangrail, Coffee, CommonsDelinker, Corpx, Courcelles, Cretog8, Cwgordon7, D. Cerv, DCEdwards1966, DL81292, DancingPenguin, Darth Panda, Davidcannon, Dbfirs, Dekisugi, Desiphral, Diaboli, Docu, Donarreiskoffer, DoorFrame, Dragoon17, Drmgrvy, Dusti, Dysprosia, E Pluribus Anthony, ERobson, Eagle4000, EdGl, Edgar181, Edward, Eeyore tim, Egg Centric, Eggman64, Eleassar777, Elekhh, Elmsat, Elockid, Eopsid, Epbr123, Eric Sandholm, EugeneZelenko, Excirial, Extra999, Eyedubya, Eyeofhorace, Farosdaughter, Figureskatingfan, Filemon, Francium12, Frozen4322, Func, GFW82, Garion96, Gcorral, GeeJo, Geremia, Gergis, Ghaly, Gidonb, Ginsengbomb, Glen, Gnosis1185, Gog, Gparalikidis, Graculus, Grafen, GrayFullbuster, Gryffindor, Gwabell, HGB, HalfShadow, Hallows AG, Hamedog, Helix pomatia, HenryLi, Hermitage17, Heron, Hitman012, Hoplophile, HornetMike, Hsm rocks ya right, Hypertall, Hégésippe Cormier, INeedANameThatIsFunny, Ibbn, Introvert, Iridescent, Island, J.delanoy, JNW, JUSTM3-X, Jacob.jose, Jake Spooky, Jakeburberry, JavierMC, Jaw959, Jdeamerr, Jds10912, Jeff G., Jim457, Joblio, JohnDoe0007, Johnuniq, Jonkerz, Jorgeposada14, Joseph Philipsson, Joseph Solis in Australia, Josh howson, Josisb, Jpk, Jprksh, Jrein59, Kaare, Karl-Henner, Karok, Keenan Pepper, Keilana, KenFehling, Ketiltrout, Kevin Gorman, Klparrot, KnowledgeOfSelf, Kr4ft, Kransky, Kurieeto, Kyle1278, L Kensington, Lakers, Latebird, LcawteHuggle, LeaveSleaves, Leks81, Leszek Jańczuk, Levineps, Liftarn, Lightmouse, Logose, Lotu, Lucas175, Lugia2453, LukeStewart, Luxborealis, Lytsio, MER-C, MJCdetroit, MaNeMeBasat, Mack2, Malcolmxl5, MapsMan, Marc Lacoste, Marek69, Mark Arsten, Mark J, Martial75, Materialscientist, Matthewcgirling, Maximus Rex, McSly, Melcombe, Mgnbar, MidgleyDJ, Mightymights, Miguel.mateo, Mindmatrix, Mintleaf, Missmarple, Moreschi, Motoxrampage101, Mrbuddhafreak, Mschlindwein, Mxn, NLikhov, NTK, Nagy, Nakon, Naval Scene, Neko-chan, Nentrex, Nepenthes, NewEnglandYankee, Newbyguesses, Nick C, Noah Salzman, Nunquam Dormio, Nyttend, Olly150, Olsonnnnn, Ongar the World-Weary, Oreo Priest, Orestimba49, PBP, Panicburststs, Patrick, Pax:Vobiscum, Pearle, Pengo, Pgan002, Phaust, PhilKnight, Philip Trueman, Physchim62, Piano non troppo, PierreAbbat, Pigman, Pinethicket, Piotrus, Pleasetouchme, Popsracer, Professionalsmartpeople1, Pumpmeup, Quasipalm, R Lowry, R'n'B, RLeopard, Radishes, Readyhairman, Red Slash, Remuel, Rev. John, ULC, RexNL, Rgdboer, Riana, Richard Warrington, Rmhermen, Romanm, Ronhjones, RoyBoy, Rushbugled13, Ryaelt, Ryan Valdés, SDC, SFK2, Samrager, SarekOfVulcan, Sarenne, Scohoust, Shizhao, Simmania, Simonkeuler.1245, Sin Harvest, Skinnyweed, SmthManly, Snowolf, Snowolfd4, Sonicyouth86, Spencer, Srasku, StaticGull, Ste, Steven Zhang, Steveprutz, Stormie, Student7, Sumahoy, Sunoco, Superm401, Surfbear335, Symode09, TGalt, TUKAKKAR, Tabitcorne2, TaintedMustard, TakuyaMurata, Tangotango, Tanzy97, Taykp, Tbhotch, Tellyaddict, Terence, The Angel of Islington, The Anome, The Determinator, The Hybrid, The Thing That Should Not Be, Thecaeser, Thehelpfulone, Thekohser, Theopolisme, Thermalite, Thucydides of Thrace, Thunder8, Tide rolls, Timeshifter, Tktktk, Tlc61, Tobias Galtieri, Tom harrison, Tomatoman180, Topbanana, Torchwoodwho, Tradewater, Tree Kittens, Trengarasu, Triesault, Tsxado, Tuanglen, Una Smith, Uncle Dick, Uuuussssseeeerrrrnnnnaaaammmmeeee,

Vegard20, Violetriga, Vollex, Vranak, Waggers, Waldir, Wangi, Wasbeer, Wavelength, Waywardhorizons, WeißNix, Wereon, Will Beback, Wilsbadkarma, Wimt, Yellowtruck, Ylem, Yngvadottir, Zanimum, Zap Rowsdower, Zfr, Zzuuzz, Александър, अीष भटनागर, 718 anonymous edits

Race_and_ethnicity_in_the_United_States_Census *Source*: http://en.wikipedia.org/w/index.php?title=Race_and_ethnicity_in_the_United_States_Census *Contributors*: (jarbarf), 10metreh, 11dkaelin, 18alex12, 18ppincock, 1exec1, 814hgc, ABF, AVand, Abdalla A, Abductive, Abeg92, Acather96, Access Denied, Aesopos, Afishslappedme, Agaccen, Ahoerstemeier, Aitias, AjaxSmack, Alajar.pds, Alan Liefting, Alansohn, Alexf, Alexius08, AlexiusHoratius, Allenmcbrewer, Alpha Quadrant (alt), Altenmann, Altynsky1234, Anaxial, AnkhMorpork, Anna Lincoln, Annebanan, AnonMoos, Anotherkid01, Antandrus, Argentina 678, Arnaudh, Arthena, Astrocommander, Atari400, Avengah, Avicennasis, BGManofID, Backie54, Barticus88, Bastique, Bava Mckenzie, Beastinbman, Bender235, Bgwhite, Big Adamsky, BigPimpinBrah, BigTom72, Bigjesus69, Billysueshitdick, Binksternet, Birdman1, BlackMamba24601, Blue520, Blueeli7, Bluefist, Bnet504, Bobianite, Bobiscool123, Bobo192, Boing! said Zebedee, Bongwarrior, Boongie, Boy2013, BrainyBroad, BrendelSignature, Brian Crawford, BrndnHD, Bub0297, Builtseven, Bulldog 123 004, C.Kent87, C628, CATDADDY103, CJLL Wright, CMW275, CRAFT910, CactusWriter, Calagirly678, Cali567, Camlovesjanet, CanadianLinuxUser, CardinalDan, Carmzmelo, Carseblawlz, CartelCacique, Casey14, Cesargevert, Chinggy332, ChompyBrock, Chris the speller, ChrisHodgesUK, Chrispalmer97, Chrisroe51, Citynoise, Ckk23, ClairSamoht, Claviere, Cleared as filed, Closedmouth, Coffee, Cold Gauk, Colorfulharp233, Cometstyles, Coolnessmonster, Cosmo77, CrazyTalk, Crazyman08, Cretog8, Cst17, Curiousier, Curps, DCGeist, DHN, DO'Neil, DVdm, DWoodHenk, Dac04, Damiencutajar1234567890, Dark Tea, Dark Tichondrias, DarkAudit, Darktower 12345, Daven200520, Davidpeterson666, Davo0777, Dbachmann, Dbiel, DeadEyeArrow, Denise Oliver-Velez, Dethme0w, Dfrg.msc, Diannaa, Didhekno, Discospinster, Distal24, Djramone, Dmitri Lytov, DopefishJustin, DouglasCalvert, Doulos Christos, Download, Dr.PeterRobertson, Drmies, Drumguy8800, Duanedonecker, E2eamon, EDGARR, EWikist, Edderso, Eden1D, Edison, EditingBeast97, Edward, ElTyrant, Elockid, Eminemisthebestmcever, Emmys7475, EnemyOfTheState, Epbr123, Eregli bob, EronMain, Erudy, Euchiasmus, Ewanwake, Excirial, FSG, FUQFUQFUQFUQFUQ, Faggtoid123, Fagmaster1339, Famartin, Famdopel, Farras Octara, Favonian, Fencepost, FilipeS, Firstorm, FlamingDuck, Flightx52, Floul1, Fluffernutter, FlySoHigh.6783, Fnlayson, Forever Dusk, Foxcloud, Frankie816, Frecklefoot, FrickFrack, Fritzpoll, Fuckingscumbag, Furado, Futurist110, Fæ, Gardar Rurak, Gatemansgc, Gcorral, Gertrudethetramp, Gggh, Ghfgfthf1, Ghostridr90, Gimmetrow, Glane23, Glarimer, Gniggz, GodSka, Golbez, Golfandme, Golgofrinchian, GorillaWarfare, Gouldevil, Gramaic, Gsp8181, Guanlongwucaii, Gunnybunny1337, HJ Mitchell, HalfShadow, Hammychaz, HarleyRowland, Hattar393, HeartofaDog, HeleneSylvie, Hello32020, Henri Müller, Henrybest3, Herstory12, Heymon32, Hihellowhatsup, Hiimroho, Historicalgeniues, Hmains, Hongooi, Hottentot, Howicus, Huntster, HupHollandHup, I am a freakazoid, IANVS, Ian Struan, Igetdatars, Ijon, Infodon, Intranetusa, Iridescent, Isaaccp, Ishu, Isuperjoei, Ithinkhelikesit, J.delanoy, JNW, JWB, JaGa, Jaggersmum2, JamesBWatson, Jamespeterka, Jargoness, Jassim khalid mohamad, Jauche vs. Mist, Jay Litman, JayDie, Jebiah72, Jebus989, Jeff G., Jersey Devil, Jh12, Jiang, Jim1138, Jj137, Jlaw14, Jlm97jlm, Jncraton, Joe Decker, Joefromrandb, Joemsu, Joevany100, John Reid, Jonny34567, JonnyOB17, Josemamc, Joseph Solis in Australia, Josh Parris, Josh3580, Jptate12, Jshook08, Jsmlb45, Ju6613r, Juangarcia87, Juliannovarr, JulieKO, Jusdafax, JustAMuggle, Jwoodger, Jyroberson, KashMony, Kathovo, Keetoowah, Keithpenson, Kenaldinho10, Keraunos, Killtimmy99, King of the Court, Kki1990, Kku, Klilidiplomus, Kman543210, Kmoney11, KnightRider, KnowledgeOfSelf, Koavf, Koklia123, Kortiz, Kralizec!, Kross, Kukkurovaca, L Kensington, LAIRDpowers, LJL2112, Ladies.man32, Lawikitejana, Lbbull, LedgendGamer, LeeMulod333, Legos9877, Leszek Jańczuk, Liamthomas69, Lick14443, LightAnkh, Lindsayoris15, LittleOldMe, Lokiloki123, Loo742o, Lord of the Pit, LrdSlvrhnd, Luisvideogames, Lukobe, Luna Santin, Lupin, Lupita°, LuvsInfoLots, Lynne64, M5891, MER-C, MK121, MShabazz, Malik Shabazz, Marek69, Mastergamejim, Mateo SA, Materialscientist, Mato, Matterfoot, MatthewW7, Mauler90, Mcitchy, Mcontreras720372, Mediran, MelbourneStar, Melchior.meter, MementoVivere, Menchi, Mhartman4, Mikenassau, Mistakefinder, Mitch Ames, Mockko, Monk, Monkey56789, Monty845, Morning277, Mr.interneter, N5iln, Nahlaugh, NameIsRon, Naraht, NawlinWiki, NeonMerlin, Neutrality, New Age Retro Hippie, NewEnglandYankee, Nickking3, Nneonneo, Nnmmdevelopment, Noahclub, Ntsimp, Nummer29, Oddbodz, Ohnoitsjamie, Ojay123, Olivier, Omdfw, One-big-ball, Onnnnnnnnn9, Orlady, Oscar the dinosaur, Ourtragichero, Padishah5000, Palfrey, ParisianBlade, Parkwells, Patrick, Peak, Pejman, Pennypup111, Perfect Proposal, Peteforsyth, Philg88, Philip Trueman, Pianomovies, PinchasC, Pk041, Polytetrahadron, Poopface002, Professor marginalia, Prophetess mar, Pschemp, Pureaswater, Quackslikeaduck, Qwerty12469, R9tgokunks, RA0808, Rachellovesyou188, Ragingscythe, Ram-Man, Raretomediumrare, Razor2988, Recognizance, RedWolf, Redaktoer, Regancy42, Reinyday, Rexwarrior, Rey2k7, Rhythmnation2004, Rich Farmbrough, Riley Huntley, Riptoo, Rivertorch, Rjensen, Rjwilmsi, RobertG, Robertoliver69, Rocketrod1960, Rogerio p Furtado, Roke, Ronhjones, Rrburke, Rror, Rsrikanth05, Ruakh, Rugadh, Rushbugled13, Ruy Pugliesi, Rwoods3, Ryan032, Sadpandaa, SamEV, Samsta3102, Sanfranman59, Satellizer, Savidan, Scarletletter4455, SchfiftyThree, Scorpios, ScottyBerg, Scoutersig, Seany217, Seaphoto, Sertyu222, Shadowjams, Shadowlynk, SheffieldWikimapian, Shell Kinney, Shithead456, Shreevatsa, Sicilianmandolin, Sillybillypiggy, Simimyles, SimonP, Sinhakamalkant, Sir II. Unknown, Skatingartist, SkerHawx, Slabodsky, Slon02, Smelliestdooever, Some jerk on the Internet, Sophie, Soupforone, SouthernComfort, SouthernNights, Squrit119, Stevey7788, Stok3z, Supercalifrag778, Supportobama1, Swagswagswagswagswag, Szyslak, T0134125H, TANKKER5, Taztouz1, Tbhotch, Tcncv, Tedickey, Telekenesis, Terces.natas.skcus, Terrafire, TerrorSonghai, Thane, The Copyeditor, The Nut, The Ogre, The Thing That Should Not Be, The voice of mud, TheKid188810, TheSuave, Thefunk909, Thegreyanomaly, Theopolisme, Thescrambles2, Thingg, Tide rolls, Tifdunc, Toby Bartels, Tolly4bolly, Tony Fox, TooPotato, TotallyAfemiminist, Tow, Tremorillo, Tristt, Trlovejoy, Trololololol1, TrueComputerKing33, Tweeheart, TympanicAlloy35, Tysabeast, Ubardak, Ucucha, Ultron, Urmama6969696969, User1acount, Vbd, Vegardkild, Velella, Vostok 129, Vrenator, Vrgrandhi, Wackywace, Wayne Slam, Weedwhacker128, WhatamIdoing, Who.was.phone, Widr, WikiWikiWoo15, Wikih101, Wikipelli, Wikiscribe, Wimt, Wonderland42, Woogee, Woohookitty, WookieInHeat, Wowserver, Wreyufg, Wtmitchell, Wuddawuddawudda, Xcaceres, Xenophrenic, Yahel Guhan, Yamamoto Ichiro, Yardserea, Yerevanci, Yooojasmane, Yopohari, Ysabellabug, Yunshui, Zgarbis, Zntrip, 964 anonymous edits

Per_capita_income *Source*: http://en.wikipedia.org/w/index.php?title=Per_capita_income *Contributors*: 0, 198.207.223.xxx, Accurizer, Aelius28, Ahoerstemeier, Ahuskay, Altaïr, Anir1uph, Antandrus, Arsenalfc06, Arthur Rubin, Ausir, Austinrok, Avenged Eightfold, BarretB, Bart133, Beland, Bhny, Black men in pants, Blue520, Bobo192, BrendelSignature, Brian0918, Brianhe, Brobdingag, Brockert, Bvu wieuwi, Calmer Waters, Canderson7, Carlosguitar, CarminPolitano, Cdc, Cheque some, Christian75, CieloEstrellado, Clearcontent, Cocoma, Curps, Cypoet, DARTH SIDIOUS 2, DFGGH, DJ Clayworth, DL5MDA, DO'Neil, DVD R W, Dalillama, Dark567, Dave6, DavidGC, Ddlamb, Deetdeet, Dekisugi, Denisarona, Driftwoodzebulin, Dtellett, DuRoad, Duoduoduo, Dustinkelley, Eastcote, Eastlaw, Elmondo21st, Elockid, EncMstr, Enviropearson, Epbr123, EurekaLott, Everyking, Evil saltine, Fingers-of-Pyrex, Foobar35, Fraggle81, Fram, Friosurely, GateKeeper, Gene.arboit, Giftlite, Girishsharma86, Gmaxwell, Graham87, Green Giant, Gtstricky, HIDECCHI001, Harlem Baker Hughes, Hgb1217, Ifnord, Ilikeverin, Imakegoodedits, Indon, Ixfd64, J.delanoy, JForget, JFreeman, JYolkowski, James086, JamieA, Jaranda, Jmarchn, Jogloran, John, Joseph Solis in Australia, Kkm010, LAX, Leslie Mateus, Leszek Jańczuk, Lfstevens, Longhair, Lotje, LuK3, Lunarune, MaCRoEco, Maurreen, Mike Payne, Miketwo, Morgan Leigh, Mxn, Myanw, Neechalkaran, Newkai, Nijusby, Nimur, Niteowlneils, Nuberger13, Pax:Vobiscum, Pdcook, Philip Trueman, Piano non troppo, Pinethicket, Polylepsis, Prashanthns, Priyatu, Quanneur, QuixoticKate, RachelleLin, Raymondwinn, Rebel, Reggionh, Resurgent insurgent, RexNL, Robtim86, Rocky Young Crobee, Rush Luagh, SLi, Shadowjams, Shanemcd, Simeondahl, SimonP, Snoyes, Spencerthomas55, Spiffy sperry, Sullykid77, Tagishsimon, TakuyaMurata, Tanvir Ahmmed, TastyPoutine, Taxman, Teapot207, Tekana, TheMidnighters, Thebestofall007, Thingg, Tim Ivorson, TimShell, Titodutta, Tobias Hoevekamp, Toffee7053, Travelbird, Van helsing, Vary, VerruckteDan, Viridian, Volgar, Voyevoda, Walton One, Wekrjlsdlklksdrujl, Wiki alf, William Avery, Wimt, Wmahan, Zeppelin462, خالقداد, 362 anonymous edits

Poverty_threshold *Source*: http://en.wikipedia.org/w/index.php?title=Poverty_threshold *Contributors*: Aaboltaabol, Acroterion, Aengus, AeriDenise, Agradman, Ahackworth, Aillema, Aleksd, Aliyevramin, All Is One, Allens, Amazingcreator79, Andy Xucha, Anomalocaris, Antandrus, Arctic Kangaroo, Arthena, Astronautics, AuburnPilot, Ausir, Barek, Beland, Bigpapi6292, Bluerasberry, Bookandcoffee, BrendelSignature, Burrburr, Byelf2007, CLW, Can't sleep, clown will eat me, Cantus, Cardsplayer4life, Cassivs, Cconnett, Cflm001, Chozanm, Chuckdevore, Cielomobile, Cirino73, Ckatz, Classicrockfan42, Closenplay, Cntras, Cocoaguy, Coelhodg, Commander Keane, Common Man, Cretog8, Cuvtixo, Dambrosio, Damicatz, DanKeshet, David Biddulph, DevilEditerLol, Dgies, Dismas, Dlohcierekim, Drmies, Duoduoduo, Dwrcan, Dystopos, Ed Poor, Edgar181, Edward, Egoadvocate, ElentariAchaea, ElinorD, Elitehobo, Epbr123, Eraserhead1, Eurosong, Fakelvis, Falconleaf, Fasttwitch, Fconaway, Fidalgo, Fitzaubrey, Foober, Frenzie, Gaff, Galf, Gareth Griffith-Jones, Gavin.perch, Gene.arboit, Gilliam, Gizziiusa, Globalsolidarity, Goldom, Grick, Guoguo12, Hadrianheugh, Hamedy1986, Haoie, HarveyA, Hector.C.Jorge, Hede2000, Henrygb, Heron, HexaChord, Hibernian, Hrh80, Human.v2.0, Humanflux, Hurutch, Ian13, Ivan1984, Iviney, J.delanoy, JakubATusaDotcom, Janshouse, JavaAndCSharp, Javadane, JeremyA, Jhortman, Jim1138, Jimtaip, Jitse Niesen, Joseph Solis in Australia, Josh Parris, Joy, Junling, Kanook, KathrynLybarger, Kellen, Kelly Martin, Ken@popp2.com, Khanra, Kirachinmoku, Kissmequick, KungFuMonkey, Lalit Jagannath, Lawrencekhoo, Leslie Mateus, Leszek Jańczuk, Levineps, LilHelpa, LilyKitty, Logical Fuzz, LolChavista, Lost tourist, MOA12345, Maralia, Marek69, Materialscientist, Mattbrundage, Matthewcgirling, Maurreen, Maxis ftw, McDutchie, Metta Bubble, Million Moments, Minimac, Misanthr0pe, Monsday, Mrshannigan, Mshecket, Muhandes, Musiphil, Mydogategodshat, Mynameisnotpj, Mysdaao, Möchtegern, Nach0king, Nagy, Naniwako, Naval Scene, Neptunius, Niceguyedc, NigelR, Nihiltres, Nishkid64, Noodleki, Noozgroop, Oceanflynn, Oledr, Otolemur crassicaudatus, P.bertazzi, Paddles, Passionless, Philip Trueman, Phoebe, Piotrus, Policy Person, Poopeater111, Postmortemjapan, Professortimithy, Quest for Truth, QuiteUnusual, R1000R1000, RJ010890, RainbowOfLight, ReiVaX, Revth, RexNL, Riana, Rich Farmbrough, Rjwilmsi, RobyWayne, Roke, SDnogo, Sam Hocevar, Sammo, Sapphic, Sarahj2107, Scarsxrunxdeep, SeanAhern, Search4Lancer, Septegram, Sh-otaibi, Shannon bohle, Sharkface217, Sharnak, Sheetal.sehgal, Shoy, SiobhanHansa, Skysmith, Slhurst, Smallman12q, Smokecrack, Snipson23, Squashy, Sreenand wiki, Staszek Lem, Strait, Tanaats, Team9a, Terjepetersen, The Mark of the Beast, The man stephen, Thingg, Tigga, Timc, Timwi, Tinton5, Toh, Tommy2010, TouristPhilosopher, Twang, Ultramarine, Valdomiro, Vanished user uih38riiw4hjlsd, Virtakuono, Volunteer Marek, Vungadu, WAGATHEE, WhatamIdoing, Wing Nut, Woutersmet, Wtmitchell, Yaktx, Yoyoyoitsrob, Zappa711, Zouf, 378 anonymous edits

Time_zone *Source*: http://en.wikipedia.org/w/index.php?title=Time_zone *Contributors*: *drew, 159753, A. B., A.Ou, A0404141041, A8UDI, AHeneen, ANDREJACK, Aaaallleex, Aaronw, Absalom89, Acegikmo1, Acs4b, AfDproXX, Ahoerstemeier, Alanmak, Alansohn, Albmont, Alex Bakharev, Alksentrs, AllyUnion, Alynna Kasmira, Ambarish, Aminullah, Amit6, Amolkold, Amorim Parga, Amorymeltzer, Andre Engels, Andrewpmk, Androl, Anneman, Antandrus, Arichnad, Arthena, Ashley Y, Asterion, Australia boy, AxelBoldt, BD2412, BIL, Banhtrung1, Basil.bourque, Bastin, Bazonka, Bdodo1992, Bearcat, Beland, Bellczar, Ben Arnold, Bentogoa, Bermlee, Bethpage89, Big Adamsky, Billybob112, Bjankuloski06en, Bjarki S, Bjcairns, Blahma, Bluelion, Bobo192, Bomac, Bongwarrior, Breno, Brian Kendig, Brian Patrie, Bryan Derksen, Budlight13, Buggery Boar, Bunthorne, Bushcarrot, CQ, Caese, Calor, Calton, CambridgeBayWeather, Can't sleep, clown will eat me, Canjth, CaribDigita, CatherineMunro, Celarnor, Cenarium, Centpacrr, CesarB, Charlene.fic, Chihuahua State, Chipmunkdavis, Chmod007, Chochopk, Chris24, ChrisGualtieri, Christy747, Chuck SMITH, Chun-hian, Chuq, Chuunen Baka, Circeus, Cirt, Classical geographer, Cliffb, Closedmouth, Cnmuk, Coder Dan, ColinKennedy, Colonel Cow, Commdor, CommonsDelinker, Comrade Graham, Confer44, Conscious, Conversion script, CopperSquare, Corumplex, Crd721, Cst17, Curps, Cyrus the virus grissimo, D-Rock, D15724C710N, DARTH SIDIOUS 2, DMarvin, DO'Neil, DPH1110, DVD R W, Da Vynci, Dale Arnett, Damian Yerrick, Damirgraffiti, Darth Panda, David Biddulph, DavidGrayson, Davidcannon, Davodd, Dcsohl, Deelkar, Denelson83, Derek Ross, Derild4921, Disill, Diti, Djszucs, Dmacgr 22, Dmeranda, Dmh, Dmn, DocWatson42, Docu, Donarreiskoffer, Doniago, Donner60, DoubleBlue, DouglasCalvert, Download, DrBizarro, Dragon taker, Drbreznjev, Drbug, Dreadstar, Drh104, Dsyzdek, Duncan, Duncan7670, DyNama, Dysepsion, E. Ripley, ESkog, EamonnPKeane, Earl Andrew, Earthlyreason, Ebelular, Ed Poor, Edward, Eequor, Eerlend, Ehrenkater, Ekren, Enchanter, Eomund, Eoujw, Epbr123, Eraserhead1, Erasmussen, Eric Herriman,

Eric-Wester, Eric119, Esmito, Esperant, Eubulides, Euchiasmus, Evice, Evil Monkey, Evil saltine, Ewawer, Excirial, Ezhiki, Ezhuttukari, F. F. Fjodor, FB 2000, FF2010, Facts707, Fanyavizuri, Farosdaughter, Farzanegan, FayssalF, Federico Tortorelli, Feedmecereal, Feezo, Feinoha, Fennec, Ferkelparade, Fffdervis, Fieldday-sunday, Fmph, Fonadier, Forseti, FrancoGG, Freddiecaveman, FreplySpang, Fresheneesz, Funandtrvl, Fundamental metric tensor, Fuzheado, Gabbe, Ganteng, Garfield226, Gary King, Gerry Ashton, Giant89, Giftlite, Glane23, Glenn, Glenn L, Gogo Dodo, Golbez, Goobergunch, GorillaWarfare, Gorrister, Goudzovski, Graeme Bartlett, Grafen, Graham87, Grassfire, Grick, Groogle, Ground Zero, Grstain, Guaka, Gzornenplatz, H2g2bob, HJ Mitchell, Hadal, Hadiyazzi, Haham hanuka, Haihe, Haikupoet, Hajor, Happening1, Hapsiainen, Harp, Harriv, Hazhk, Heb, Heitordp, Hellerick, Henrygb, Hesperian, Hi is bob, HistoryAddict, HortusBlizzard, Htmlland, Humilulo, HumphreyW, Hvn0413, Hydrargyrum, IKidaround, IMSoP, IQveteran, Ido50, Ignis Fatuus, IgorMagic, Imoen, Imz, Incnis Mrsi, Indefatigable, Iner22, IngSoc BigBrother, Instantnood, Intgr, Iridescent, Island, Itai, Itisonly4me, Ixfd64, J-stan, J.delanoy, J.smith, JHunterJ, JLaTondre, JNW, JackofOz, Jakes18, Jamadagni, Jarrodiscool141314, JavierMC, Jc3s5h, Jcra, Jd4508, Jda, Jeff G., Jeff3000, Jeffq, Jerome Charles Potts, Jerzy, Jesper Norgaard, Jiang, Jim1138, Jimp, Jj137, Jla, Jnestorius, Joe Kress, JoeSmack, Joey2cool2007, John Paul Parks, John Quincy Adding Machine, John of Reading, JohnOwens, JohnWittle, Jojhutton, Jojit fb, JonathanJosephsen, Jonathunder, Joostjodel, Joseph Solis in Australia, Jpk, Jplimpz, JulDes, Juliancolton, Julianp, Jumbuck, Jusjih, Just H, Jérôme, Karl Palmen, Karl-Henner, KathrynLybarger, KayEss, Ke4roh, Keeves, KeithH, Kh2mohamed, King of Hearts, Kingal86, Kiwi128, Klemen Kocjancic, Knowledge Seeker, Krawunsel, Kubigula, Kukini, Kungfuadam, Kyle Barbour, Kylekieran, LOL, Lacrimosus, Lalah1294, Lambiam, Lambyte, Lamrongpeng5, Last Avenue, Lee J Haywood, LeeTraynor, Leeyc0, LeinsterLion, Leucius, Leuko, Lewisjon, LilHelpa, Lincmad, Lir, Lizzie Harrison, Llywrch, Logan, Lolm8, LorenzoB, Lotje, Lucamauri, Lugia2453, Luis Dantas, Luna Santin, M, MKar, Mac, MacTire02, Macaddct1984, Magnus Manske, Maksim L., Malrase, Mancunius, Marechal Ney, Marek69, Mariusmiti, Mark Leander, MarkSweep, Marshman, Martarius, Mathew12123434, MatthewVanitas, Mav, MaxDZ8, Mayfoev, Mayooresan, McSly, Mean as custard, Meelar, Meini, MelbourneStar, Mernen, Mert34, Meters, Michael Glass, Michael Hardy, MichaelBillington, Michfan2123, Midnightblueowl, Mild Bill Hiccup, Minipie8, Minna Sora no Shita, Miquonranger03, Miranda, Miros 0571, Mjs1991, Modulatum, Mojo-chan, Monkeyman, Monni95, Monoxide333, Montrealais, Moocha, MoraSique, Morga555, Mormegil, Mr Bartels, Mr. Ajax, MrWolfy, Mrk695, Mryg, Mwarren us, Mxn, N1ywb, Naive cynic, Nanshu, Natpowning, Naval Scene, NawlinWiki, Nealmcb, NellieBly, Neptune777, NewEnglandYankee, Nichalp, Nicholas Weiner, Nicolas Love, Nightstallion, Niteowlneils, Nitsansh, Nk, Northumbrian, Nortonius, Novalis, Numbo3, Numen, Nunh-huh, Octahedron80, Ohnoitsjamie, Oleksiy.n, Olivier, OpenFuture, Optimist on the run, OverlordQ, OwenX, Oxymoron83, P. S. F. Freitas, PDucks, Paddu, Pak21, Panos84, Paranoid, Pasquale, Patach, Patcs0013, Patrick, Paul August, Paul Koning, Paul Murray, Paullees, Pavel Vozenilek, Pejman47, Peregrine981, Persian Poet Gal, Peter Wisner, PeterThoeny, Petrig, Pgan002, Phearson, Philip Stevens, Philip Trueman, Phoenix B 1of3, Pi zero, Piano non troppo, Picapica, Picaroon, PierreAbbat, PimRijkee, Pm5, Pokrajac, Popsracer, Possum, Preslethe, Psb777, Psdekloe, Pseudonymous Rex, Pytom, Quadell, Quasipalm, Rabbit233224, Radiojon, RainbowOfLight, Randyc, Raymondwinn, Rchandra, Rclocher3, Reaper Eternal, Redrose64, Redwood57, Reisio, Renetus, RenniePet, Requestion, RexNL, Rhobite, Riana, Rich Farmbrough, Rjwilmsi, Rmallins, Rmhermen, Robsinden, Roland2, Romanm, Rory096, Royaume du Maroc, Rsrikanth05, Rt66lt, Russell E, S.Örvarr.S, SGBailey, SL93, SV Resolution, Sahgal, Samuel, Samw, Saros136, Sceptre, SchmuckyTheCat, Schneelocke, Schoen, Scipius, Scoelho86, Scott Roy Atwood, Scott3, Seba5618, Secfan, SeoMac, SephChicago, Sephiroth BCR, Sesu Prime, Seth Ilys, Sewing, Sheridan A.M. Hamilton, Shoefly, Sietse Snel, Simon J Kissane, Simon_J_Kissane, Simpsnut14, Siqbal, Sjorford, Skater, Skizzik, Skyman123, Skysmurf, Skywayman, Slawojarek, Smalljim, Smart77, Smithcraft36, Smjg, Smoken Flames, Snowdog, Sommerfeld, Sonninkikkeli, SouthernNights, Sp33dyphil, Spesh531, Spitfire, Squiffycat, Steel, Stephan Leeds, Stephen B Streater, Stephen.frede, SteveMX, Sunborn, Susurrus, Sverdrup, Swid, Swpb, T-resh, T.c.w7468, TGC55, TPK, TUF-KAT, TZ master, Tad Lincoln, Takita787, TakuyaMurata, Tarquin, Tashikelzang, Tbhotch, Tebloraf, Techman224, Techtonik, Teemu Leisti, Telempe, Tfts, That-Vela-Fella, The Duke of Waltham, The High Fin Sperm Whale, The Rambling Man, The Storm Surfer, The Thing That Should Not Be, The Tom, TheEgyptian, Thf1977, Thingg, Think outside the box, Thryduulf, Thumperward, Tide rolls, Tim Parenti, Time in Russia, TimeCurrency, TimeXCode, TimeZonesBoy, Timir2, Timneu22, Timwi, Tivolives, Tmangray, Tmol42, Tobias Conradi, Tomica, Topdeck, TopoChecker, Topory, Torpov, Transisto, TruPepitoM, TrueMahn, Tsja, Tutsuro, UhMaZe xpeK, Ultratomio, Uncle Scrooge, Undescribed, Uriber, Ute in DC, VampWillow, Vanished User 1004, Vanjagenije, Vaughan Pratt, Vee power, Velho, Versus22, Viriditas, Vsmith, Walshie79, Wangi, Watson24hour, Wavelength, WeißNix, Weregerbil, West.andrew.g, Wexcan, Widr, Wik, Wiki Wikardo, Wikidenizen, Wikilackey, Wikipediman, Will Beback Auto, Willking1979, Wissons, Wizard04, Woodstone, Woohookitty, Worldtimezone, Writ Keeper, Writherthebest100, Wrtiede, Wtmitchell, Wwutx, Xeryus, Ximi12, Xpi6, Xxeric thomasxx, Xxxx693, Xzqx, Y0u, YUL89YYZ, Yamara, Ybbor, YellowMonkey, Yosef1987, Yt788, Zakuragi, Zash20, Zidonuke, Zkzkz, Zoicon5, ZooFari, Zundark, Zzuuzz, Σ, Владимир Шеляпин, תלהק, 1556 anonymous edits

Image Sources, Licenses and Contributors

Image:Bullittpopdec.GIF *Source*: http://en.wikipedia.org/w/index.php?title=File:Bullittpopdec.GIF *License*: unknown *Contributors*:

Image:Ups 757.jpg *Source*: http://en.wikipedia.org/w/index.php?title=File:Ups_757.jpg *License*: unknown *Contributors*: David Harpe

File:Flag of the United States.svg *Source*: http://en.wikipedia.org/w/index.php?title=File:Flag_of_the_United_States.svg *License*: unknown *Contributors*: Anomie

File:Great Seal of the United States (obverse).svg *Source*: http://en.wikipedia.org/w/index.php?title=File:Great_Seal_of_the_United_States_(obverse).svg *License*: unknown *Contributors*: U.S. Government

File:USA orthographic.svg *Source*: http://en.wikipedia.org/w/index.php?title=File:USA_orthographic.svg *License*: unknown *Contributors*: User:Addicted04

File:Increase2.svg *Source*: http://en.wikipedia.org/w/index.php?title=File:Increase2.svg *License*: unknown *Contributors*: Sarang

File:Indians giving a talk to Bouquet.jpg *Source*: http://en.wikipedia.org/w/index.php?title=File:Indians_giving_a_talk_to_Bouquet.jpg *License*: unknown *Contributors*: Dantadd, Hsarrazin, Kevin Myers, Origamiemensch, Shakko, Uyvsdi, Wst

File:The Mayflower Compact 1620 cph.3g07155.jpg *Source*: http://en.wikipedia.org/w/index.php?title=File:The_Mayflower_Compact_1620_cph.3g07155.jpg *License*: unknown *Contributors*: Howcheng

File:Treaty of Penn with Indians by Benjamin West.jpg *Source*: http://en.wikipedia.org/w/index.php?title=File:Treaty_of_Penn_with_Indians_by_Benjamin_West.jpg *License*: unknown *Contributors*: Auntof6, Bukk, Frank C. Müller, Goldfritha, Hsarrazin, Man vyi, Nonenmac, Origamiemensch, Presidentman, Rlbberlin, Shakko, Thuresson, Uyvsdi, Wmpearl, 3 anonymous edits

Image:Declaration independence.jpg *Source*: http://en.wikipedia.org/w/index.php?title=File:Declaration_independence.jpg *License*: unknown *Contributors*: Aavindraa, Amandajm, Americophile, Beyond My Ken, Bohème, Boo-Boo Baroo, Cambalachero, Editor at Large, Elemaki, Fred J, Geni, GrawpSock, Harpsichord246, Hluup, Ibn Battuta, Mhby87, Misogi, Morgan Riley, Mutter Erde, Nonenmac, Panoptik, Patstuart, Piledhigheranddeeper, Pmlineditor, Savh, UpstateNYer, WTCA, Wmpearl, Wst, Xavigivax, Zolo, 20 anonymous edits

Image:U.S. Territorial Acquisitions.png *Source*: http://en.wikipedia.org/w/index.php?title=File:U.S._Territorial_Acquisitions.png *License*: unknown *Contributors*: United States federal government (converted it from JPEG to PNG and retouched it)

File:Battle of Gettysburg, by Currier and Ives.png *Source*: http://en.wikipedia.org/w/index.php?title=File:Battle_of_Gettysburg,_by_Currier_and_Ives.png *License*: unknown *Contributors*: Beyond silence, Churchh, Jweiss11, Kyro, Morgan Riley, Mtsmallwood, Möchtegern, PhilFree, Shizhao, Thuresson, UpstateNYer, Veikk0.ma, 18 anonymous edits

Image:Ellis island 1902.jpg *Source*: http://en.wikipedia.org/w/index.php?title=File:Ellis_island_1902.jpg *License*: unknown *Contributors*: AnRo0002, Aude, Infrogmation, Kjetil r, Philpill691, UpstateNYer, 9 anonymous edits

Image:Into the Jaws of Death 23-0455M edit.jpg *Source*: http://en.wikipedia.org/w/index.php?title=File:Into_the_Jaws_of_Death_23-0455M_edit.jpg *License*: unknown *Contributors*: Alvesgaspar, Apalsola, Armbrust, Berrucomons, Bjankuloski06en, Catsmeat, Christoph Braun, Esemono, Mogelzahn, Moroboshi, Nnemo, Noritaka666, Shizhao, Svajcr, Thierry Caro, Verdy p, Viloris

Image:Abernathy Children on front line leading the SELMA TO MONTGOMERY MARCH for the RIGHT TO VOTE.JPG *Source*: http://en.wikipedia.org/w/index.php?title=File:Abernathy_Children_on_front_line_leading_the_SELMA_TO_MONTGOMERY_MARCH_for_the_RIGHT_TO_VOTE.JPG *License*: unknown *Contributors*: Abernathy Family

Image:DF-ST-91-01394.JPEG *Source*: http://en.wikipedia.org/w/index.php?title=File:DF-ST-91-01394.JPEG *License*: unknown *Contributors*: DoD photo by: STAFF SERGEANT F. LEE COCKRAN

File:UA Flight 175 hits WTC south tower 9-11 edit.jpeg *Source*: http://en.wikipedia.org/w/index.php?title=File:UA_Flight_175_hits_WTC_south_tower_9-11_edit.jpeg *License*: unknown *Contributors*: User:UpstateNYer

File:USA-satellite.jpg *Source*: http://en.wikipedia.org/w/index.php?title=File:USA-satellite.jpg *License*: unknown *Contributors*: Original uploader was E Pluribus Anthony at en.wikipedia

File:Bald Eagle Portrait.jpg *Source*: http://en.wikipedia.org/w/index.php?title=File:Bald_Eagle_Portrait.jpg *License*: unknown *Contributors*: User:Saffron Blaze

File:Census-2000-Data-Top-US-Ancestries-by-County.svg *Source*: http://en.wikipedia.org/w/index.php?title=File:Census-2000-Data-Top-US-Ancestries-by-County.svg *License*: unknown *Contributors*: ANGELUS, Che829, Conscious, Davepape, Editor at Large, EurekaLott, Gpvos, JaumeBG, Jmabel, Kintetsubuffalo, MacedonianBoy, Man vyi, Mikael Häggström, Myself488, Nagy, Ras67, Tryphon, UpstateNYer, ZooFari, 17 anonymous edits

File:Top of Rock Cropped.jpg *Source*: http://en.wikipedia.org/w/index.php?title=File:Top_of_Rock_Cropped.jpg *License*: unknown *Contributors*: User:Dschwen

File:LA Skyline Mountains2.jpg *Source*: http://en.wikipedia.org/w/index.php?title=File:LA_Skyline_Mountains2.jpg *License*: unknown *Contributors*: User:Nserrano

Image:Skyline_from_Millennium_Park.jpg *Source*: http://en.wikipedia.org/w/index.php?title=File:Skyline_from_Millennium_Park.jpg *License*: unknown *Contributors*: User:Peterfitzgerald

File:Religions of the United States.png *Source*: http://en.wikipedia.org/w/index.php?title=File:Religions_of_the_United_States.png *License*: unknown *Contributors*: User:HaireDunya

Image:Capitol Building Full View.jpg *Source*: http://en.wikipedia.org/w/index.php?title=File:Capitol_Building_Full_View.jpg *License*: unknown *Contributors*: Noclip

Image:WhiteHouseSouthFacade.JPG *Source*: http://en.wikipedia.org/w/index.php?title=File:WhiteHouseSouthFacade.JPG *License*: unknown *Contributors*: Darwinius, Darxus, H005, HiraV, Infrogmation, J 1982, Jatkins, JenVan, Lycaon, Pi.1415926535, Redtigerxyz, TCY, UpstateNYer, 2 anonymous edits

Image:USSupremeCourtWestFacade.JPG *Source*: http://en.wikipedia.org/w/index.php?title=File:USSupremeCourtWestFacade.JPG *License*: unknown *Contributors*: User:UpstateNYer

File:Political System of the United States.svg *Source*: http://en.wikipedia.org/w/index.php?title=File:Political_System_of_the_United_States.svg *License*: unknown *Contributors*: User:111Alleskönner

Image:Map of USA with state names 2.svg *Source*: http://en.wikipedia.org/w/index.php?title=File:Map_of_USA_with_state_names_2.svg *License*: unknown *Contributors*: User:Andrew c

File:Obama meets with Congressional Leadership July 2011.jpg *Source*: http://en.wikipedia.org/w/index.php?title=File:Obama_meets_with_Congressional_Leadership_July_2011.jpg *License*: unknown *Contributors*: Antony-22

File:Hague Clinton May 14 2010 Crop.jpeg *Source*: http://en.wikipedia.org/w/index.php?title=File:Hague_Clinton_May_14_2010_Crop.jpeg *License*: unknown *Contributors*: User:UpstateNYer

File:FederalDebt1940to2012.svg *Source*: http://en.wikipedia.org/w/index.php?title=File:FederalDebt1940to2012.svg *License*: unknown *Contributors*: Mikhail Ryazanov

File:US Navy 060618-N-8492C-212 An Air Force B-2 bomber along with other aircrafts from the Air Force, Navy and Marine Corps fly over the Kitty Hawk, Ronald Reagan and Abraham Lincoln Carrier Strike groups.jpg *Source*: http://en.wikipedia.org/w/index.php?title=File:US_Navy_060618-N-8492C-212_An_Air_Force_B-2_bomber_along_with_other_aircrafts_from_the_Air_Force,_Navy_and_Marine_Corps_fly_over_the_Kitty_Ha *License*: unknown *Contributors*: Benchill, Docu, FAEP, FoolsWar, 1 anonymous edits

File:NYPD impala.JPG *Source*: http://en.wikipedia.org/w/index.php?title=File:NYPD_impala.JPG *License*: unknown *Contributors*: User:Pablo029

File:Photos NewYork1 032.jpg *Source*: http://en.wikipedia.org/w/index.php?title=File:Photos_NewYork1_032.jpg *License*: unknown *Contributors*: Basilicofresco, Beyond silence, Loranchet, Quasipalm, Urban, 1 anonymous edits

File:Median US household income.png *Source*: http://en.wikipedia.org/w/index.php?title=File:Median_US_household_income.png *License*: unknown *Contributors*: CLDunlap

File:South San Jose (crop).jpg *Source*: http://en.wikipedia.org/w/index.php?title=File:South_San_Jose_(crop).jpg *License*: unknown *Contributors*: User:Seanol

File:Map of current Interstates.svg *Source*: http://en.wikipedia.org/w/index.php?title=File:Map_of_current_Interstates.svg *License*: unknown *Contributors*: User:SPUI

File:Neil Armstrong in suit.jpg *Source*: http://en.wikipedia.org/w/index.php?title=File:Neil_Armstrong_in_suit.jpg *License*: unknown *Contributors*: User:W.wolny

File:University-of-Virginia-Rotunda.jpg *Source*: http://en.wikipedia.org/w/index.php?title=File:University-of-Virginia-Rotunda.jpg *License*: unknown *Contributors*: gargola87 on Flickr

File:Aerial of Texas Medical Center with Downtown Houston in the background.jpg *Source*: http://en.wikipedia.org/w/index.php?title=File:Aerial_of_Texas_Medical_Center_with_Downtown_Houston_in_the_background.jpg *License*: unknown *Contributors*: Agsftw, Hequals2henry

File:Liberty-statue-from-below.jpg *Source*: http://en.wikipedia.org/w/index.php?title=File:Liberty-statue-from-below.jpg *License*: unknown *Contributors*: User:Tysto

File:Mark Twain, Brady-Handy photo portrait, Feb 7, 1871, cropped.jpg *Source*: http://en.wikipedia.org/w/index.php?title=File:Mark_Twain,_Brady-Handy_photo_portrait,_Feb_7,_1871,_cropped.jpg *License*: unknown *Contributors*: Mathew Brady

File:Times Square 1-2.JPG *Source*: http://en.wikipedia.org/w/index.php?title=File:Times_Square_1-2.JPG *License*: unknown *Contributors*: Beyond My Ken, Gryffindor, Infrogmation, Jatkins, Jim.henderson, Julia W, LongLiveRock, Sdrtirs, Takabeg, UpstateNYer, 1 anonymous edits

File:Motherhood and apple pie.jpg *Source*: http://en.wikipedia.org/w/index.php?title=File:Motherhood_and_apple_pie.jpg *License*: unknown *Contributors*: Scott Bauer, USDA ARS

File:Michael Phelps 2009.jpg *Source*: http://en.wikipedia.org/w/index.php?title=File:Michael_Phelps_2009.jpg *License*: unknown *Contributors*: Marylandstater

File:Office-book.svg *Source*: http://en.wikipedia.org/w/index.php?title=File:Office-book.svg *License*: unknown *Contributors*: Chris Down/Tango project

File:Gnome-globe.svg *Source*: http://en.wikipedia.org/w/index.php?title=File:Gnome-globe.svg *License*: unknown *Contributors*: David Vignoni

file:speakerlink-new.svg *Source*: http://en.wikipedia.org/w/index.php?title=File:Speakerlink-new.svg *License*: unknown *Contributors*: User:Kelvinsong

File:Country road in Kentucky.jpg *Source*: http://en.wikipedia.org/w/index.php?title=File:Country_road_in_Kentucky.jpg *License*: unknown *Contributors*: Original uploader was Censusdata at en.wikipedia. Later version(s) were uploaded by Pd THOR at en.wikipedia.

File:Map of Kentucky NA.png *Source*: http://en.wikipedia.org/w/index.php?title=File:Map_of_Kentucky_NA.png *License*: unknown *Contributors*: Huebi, Jmabel, Zyxw

File:KYphysiography.jpg *Source*: http://en.wikipedia.org/w/index.php?title=File:KYphysiography.jpg *License*: unknown *Contributors*: Original uploader was Lamontacranston at en.wikipedia

File:View from Pine Mountain (Kentucky).jpg *Source*: http://en.wikipedia.org/w/index.php?title=File:View_from_Pine_Mountain_(Kentucky).jpg *License*: unknown *Contributors*: User:J654567

File:Kentucky horse farm.JPEG *Source*: http://en.wikipedia.org/w/index.php?title=File:Kentucky_horse_farm.JPEG *License*: unknown *Contributors*: User:Peterfitzgerald

File:Kybaldcypress.jpg *Source*: http://en.wikipedia.org/w/index.php?title=File:Kybaldcypress.jpg *License*: unknown *Contributors*: User:Southernky

File:Picture 491.jpg *Source*: http://en.wikipedia.org/w/index.php?title=File:Picture_491.jpg *License*: unknown *Contributors*: Original uploader was Censusdata at en.wikipedia

File:WaterfrontPkDwnt.jpg *Source*: http://en.wikipedia.org/w/index.php?title=File:WaterfrontPkDwnt.jpg *License*: unknown *Contributors*:

File:Half Moon, Kentucky.JPG *Source*: http://en.wikipedia.org/w/index.php?title=File:Half_Moon,_Kentucky.JPG *License*: unknown *Contributors*: Beeblebrox, DragonflySixtyseven, Mrherbalwarrior

File:LincolnBirthplaceJBM82908.jpg *Source*: http://en.wikipedia.org/w/index.php?title=File:LincolnBirthplaceJBM82908.jpg *License*: unknown *Contributors*: Censusdata at en.wikipedia. Later version was cropped by Fourthords at en.wikipedia.

File:U.S. Marine Hospital, Louisville, Kentucky.jpg *Source*: http://en.wikipedia.org/w/index.php?title=File:U.S._Marine_Hospital,_Louisville,_Kentucky.jpg *License*: unknown *Contributors*: Infrogmation, Sreejithk2000

File:William Goebel statue.png *Source*: http://en.wikipedia.org/w/index.php?title=File:William_Goebel_statue.png *License*: unknown *Contributors*: User:Acdixon

File:KY Governors Mansion.png *Source*: http://en.wikipedia.org/w/index.php?title=File:KY_Governors_Mansion.png *License*: unknown *Contributors*: Original uploader was Acdixon at en.wikipedia

File:KY State Capitol.jpg *Source*: http://en.wikipedia.org/w/index.php?title=File:KY_State_Capitol.jpg *License*: unknown *Contributors*: RXUYDC

File:KY-districts-108.JPG *Source*: http://en.wikipedia.org/w/index.php?title=File:KY-districts-108.JPG *License*: unknown *Contributors*: Original uploader was Seth Ilys at en.wikipedia

File:Kentucky population map.png *Source*: http://en.wikipedia.org/w/index.php?title=File:Kentucky_population_map.png *License*: unknown *Contributors*: Original uploader was JimIrwin at en.wikipedia

File:CollegeoftheBible-LexKY.JPG *Source*: http://en.wikipedia.org/w/index.php?title=File:CollegeoftheBible-LexKY.JPG *License*: unknown *Contributors*: unknown

File:2012 Toyota Camry SE -- 02-29-2012.JPG *Source*: http://en.wikipedia.org/w/index.php?title=File:2012_Toyota_Camry_SE_--_02-29-2012.JPG *License*: unknown *Contributors*: IFCAR

File:Ford F-150 XL SuperCrew -- 03-10-2010.jpg *Source*: http://en.wikipedia.org/w/index.php?title=File:Ford_F-150_XL_SuperCrew_--_03-10-2010.jpg *License*: unknown *Contributors*: IFCAR

File:Kentucky Route 80 in Pulaski County.jpg *Source*: http://en.wikipedia.org/w/index.php?title=File:Kentucky_Route_80_in_Pulaski_County.jpg *License*: unknown *Contributors*: Original uploader was Censusdata at en.wikipedia. Later version(s) were uploaded by Pd THOR at en.wikipedia.

File:High Bridge in Kentucky.jpg *Source*: http://en.wikipedia.org/w/index.php?title=File:High_Bridge_in_Kentucky.jpg *License*: unknown *Contributors*: User:Censusdata

File:Barge hauling coal, Louisville and Portland Canal.jpg *Source*: http://en.wikipedia.org/w/index.php?title=File:Barge_hauling_coal,_Louisville_and_Portland_Canal.jpg *License*: unknown *Contributors*: Original uploader was Censusdata at en.wikipedia

File:Kentucky counties map.png *Source*: http://en.wikipedia.org/w/index.php?title=File:Kentucky_counties_map.png *License*: unknown *Contributors*: Nyttend

File:N38316642 31380476 4196.jpg *Source*: http://en.wikipedia.org/w/index.php?title=File:N38316642_31380476_4196.jpg *License*: unknown *Contributors*: Censusdata, Firespeaker

File:LouisvilleNightSkyline2-small.jpg *Source*: http://en.wikipedia.org/w/index.php?title=File:LouisvilleNightSkyline2-small.jpg *License*: unknown *Contributors*: Original uploader was Stevietheman at en.wikipedia

File:LexingtonDowntown.JPG *Source*: http://en.wikipedia.org/w/index.php?title=File:LexingtonDowntown.JPG *License*: unknown *Contributors*: Cmadler, J654567, Jonesey, Magog the Ogre, Xnatedawgx, 3 anonymous edits

File:Covington KY.jpg *Source*: http://en.wikipedia.org/w/index.php?title=File:Covington_KY.jpg *License*: unknown *Contributors*: Michael Barera

File:William T. Young Library.jpg *Source*: http://en.wikipedia.org/w/index.php?title=File:William_T._Young_Library.jpg *License*: unknown *Contributors*: User:J654567

File:Speed Engineering School, Louisville, Kentucky (1067).jpg *Source*: http://en.wikipedia.org/w/index.php?title=File:Speed_Engineering_School,_Louisville,_Kentucky_(1067).jpg *License*: unknown *Contributors*: Taken and uploaded by Jeffrey B. Morris on 6-6-06

File:Buffalo trace distillery.jpg *Source*: http://en.wikipedia.org/w/index.php?title=File:Buffalo_trace_distillery.jpg *License*: unknown *Contributors*: Kaplansa

File:Werne's Row 4th and Hill, Old Louisville.jpg *Source*: http://en.wikipedia.org/w/index.php?title=File:Werne's_Row_4th_and_Hill,_Old_Louisville.jpg *License*: unknown *Contributors*: Original uploader was Retired username at en.wikipedia

File:U.S. 23 Country Music Highway Museum.jpg *Source*: http://en.wikipedia.org/w/index.php?title=File:U.S._23_Country_Music_Highway_Museum.jpg *License*: unknown *Contributors*: User:J654567

File:Hot Brown Kurtz.jpg *Source*: http://en.wikipedia.org/w/index.php?title=File:Hot_Brown_Kurtz.jpg *License*: unknown *Contributors*: User:Shadle

File:University of Louisville marching band, Churchill Downs Twin Spires.jpg *Source*: http://en.wikipedia.org/w/index.php?title=File:University_of_Louisville_marching_band,_Churchill_Downs_Twin_Spires.jpg *License*: unknown *Contributors*: Cnyborg, Explicit, Micki, W.marsh, Yahel Guhan

File:LouisvilleSluggerMusem.jpg *Source*: http://en.wikipedia.org/w/index.php?title=File:LouisvilleSluggerMusem.jpg *License*: unknown *Contributors*: User Derek.cashman on en.wikipedia

File:Kentucky quarter, reverse side, 2001.jpg *Source*: http://en.wikipedia.org/w/index.php?title=File:Kentucky_quarter,_reverse_side,_2001.jpg *License*: unknown *Contributors*: T. James Ferrell, employee of the US Mint

File:Thunder over louisville 2006.jpg *Source*: http://en.wikipedia.org/w/index.php?title=File:Thunder_over_louisville_2006.jpg *License*: unknown *Contributors*: Shawn Skriver

File:Horse farm in Kentucky.jpg *Source*: http://en.wikipedia.org/w/index.php?title=File:Horse_farm_in_Kentucky.jpg *License*: unknown *Contributors*: Original uploader was Censusdata at en.wikipedia

File:Daniel Boone National Forest Tater Knob.jpg *Source*: http://en.wikipedia.org/w/index.php?title=File:Daniel_Boone_National_Forest_Tater_Knob.jpg *License*: unknown *Contributors*: Original uploader was Mzzl at en.wikipedia

File:Ohio River 282.jpg *Source*: http://en.wikipedia.org/w/index.php?title=File:Ohio_River_282.jpg *License*: unknown *Contributors*: Censusdata (talk). Original uploader was Censusdata at en.wikipedia

File:Shops along Fountain Square in Bowling Green, Kentucky 2008.JPG *Source*: http://en.wikipedia.org/w/index.php?title=File:Shops_along_Fountain_Square_in_Bowling_Green,_Kentucky_2008.JPG *License*: unknown *Contributors*: OPMaster at en.wikipedia

File:Flag of California.svg *Source*: http://en.wikipedia.org/w/index.php?title=File:Flag_of_California.svg *License*: unknown *Contributors*: Devin Cook

File:Flag of Texas.svg *Source*: http://en.wikipedia.org/w/index.php?title=File:Flag_of_Texas.svg *License*: unknown *Contributors*: Abjiklam, Alkari, Anime Addict AA, Awg1010, Computron, Dbenbenn, Duduziq, Dzordzm, Fry1989, Homo lupus, Hydrargyrum, Jrooksjr, Juiced lemon, Juliancolton, Mattes, Nightstallion, Piastu, Serinde, Sterling.M.Archer, Str4nd, Wars, 7 anonymous edits

File:Flag of New York.svg *Source*: http://en.wikipedia.org/w/index.php?title=File:Flag_of_New_York.svg *License*: unknown *Contributors*: State of New York

File:Flag of Florida.svg *Source*: http://en.wikipedia.org/w/index.php?title=File:Flag_of_Florida.svg *License*: unknown *Contributors*: Awg1010, Dbenbenn, Denelson83, DenghiùComm, Dzordzm, Ebyabe, Fry1989, Himasaram, Homo lupus, Hystrix, Madden, Mattes, Nightstallion, Pumbaa80, Reach Out to the Truth, Reuvenk, Rocket000, Theo10011, Vantey, Zscout370, 13 anonymous edits

File:Flag of Illinois.svg *Source*: http://en.wikipedia.org/w/index.php?title=File:Flag_of_Illinois.svg *License*: unknown *Contributors*: Awg1010, Cycn, Denelson83, Dual Freq, Dzordzm, Fry1989, Homo lupus, Hoo man, Hydrargyrum, Kranar drogin, Mattes, Permjak, Smooth O, Svgalbertian, TheCatalyst31, Theo10011, 7 anonymous edits

File:Flag of Pennsylvania.svg *Source*: http://en.wikipedia.org/w/index.php?title=File:Flag_of_Pennsylvania.svg *License*: unknown *Contributors*: Anime Addict AA, AnonMoos, Awg1010, Denelson83, Dzordzm, F. F. Fjodor, Fry1989, Homo lupus, Mattes, Niagara, Serinde, Simtropolitan, Svencb, Svgalbertian, 4 anonymous edits

File:Flag of Ohio.svg *Source*: http://en.wikipedia.org/w/index.php?title=File:Flag_of_Ohio.svg *License*: unknown *Contributors*: User:SiBr4

File:Flag of Michigan.svg *Source*: http://en.wikipedia.org/w/index.php?title=File:Flag_of_Michigan.svg *License*: unknown *Contributors*: Awg1010, Denelson83, Dzordzm, Fry1989, Herbythyme, Homo lupus, Hoo man, Mattes, Serinde, Smooth O, Svgalbertian, Theo10011, Werewombat, 5 anonymous edits

File:Flag of New Jersey.svg *Source*: http://en.wikipedia.org/w/index.php?title=File:Flag_of_New_Jersey.svg *License*: unknown *Contributors*: User:ChrisRuvolo, User:Denelson83

File:Flag of the State of Georgia (1956-2001).svg *Source*: http://en.wikipedia.org/w/index.php?title=File:Flag_of_the_State_of_Georgia_(1956-2001).svg *License*: unknown *Contributors*: Original uploader was Zscout370 at en.wikipedia

File:Flag of North Carolina.svg *Source*: http://en.wikipedia.org/w/index.php?title=File:Flag_of_North_Carolina.svg *License*: unknown *Contributors*: Awg1010, Dbenbenn, Denelson83, Dzordzm, Fry1989, Homo lupus, Hydrargyrum, Illegitimate Barrister, Serinde, Smooth O, Svgalbertian, Vantey, Wknight94, Zscout370, 6 anonymous edits

File:Flag of Virginia.svg *Source*: http://en.wikipedia.org/w/index.php?title=File:Flag_of_Virginia.svg *License*: unknown *Contributors*: ABCD, Awg1010, Dzordzm, Eugenio Hansen, OFS, Fry1989, Homo lupus, Hoodinski, Hoshie, Juiced lemon, Kam Solusar, Kelvinc, Smooth O, Svgalbertian, 12 anonymous edits

File:Flag of Massachusetts.svg *Source*: http://en.wikipedia.org/w/index.php?title=File:Flag_of_Massachusetts.svg *License*: unknown *Contributors*: State of Massachusetts

File:Flag of Indiana.svg *Source*: http://en.wikipedia.org/w/index.php?title=File:Flag_of_Indiana.svg *License*: unknown *Contributors*: User:Denelson83

File:Flag of Washington.svg *Source*: http://en.wikipedia.org/w/index.php?title=File:Flag_of_Washington.svg *License*: unknown *Contributors*: WA Secretary of State

File:Flag of Tennessee.svg *Source*: http://en.wikipedia.org/w/index.php?title=File:Flag_of_Tennessee.svg *License*: unknown *Contributors*: (of code)

File:Flag of Missouri.svg *Source*: http://en.wikipedia.org/w/index.php?title=File:Flag_of_Missouri.svg *License*: unknown *Contributors*: Awg1010, Cycn, Denelson83, Dzordzm, Fry1989, Homo lupus, Hydrargyrum, Juiced lemon, Rainman, Robth, Smooth O, Svgalbertian, Zscout370, 1 anonymous edits

File:Flag of Wisconsin.svg *Source*: http://en.wikipedia.org/w/index.php?title=File:Flag_of_Wisconsin.svg *License*: unknown *Contributors*: Denelson83, Dual Freq, Dzordzm, Fry1989, Homo lupus, Rfc1394, Royalbroil, Smooth O, Svgalbertian, Zscout370, 1 anonymous edits

File:Flag of Maryland.svg *Source*: http://en.wikipedia.org/w/index.php?title=File:Flag_of_Maryland.svg *License*: unknown *Contributors*: Abraham, Anime Addict AA, Awg1010, Dbenbenn, Dzordzm, Fry1989, Homo lupus, Illegitimate Barrister, Juiced lemon, Nightstallion, Serinde, SiBr4, Smooth O, Zscout370, 1 anonymous edits

File:Flag of Arizona.svg *Source*: http://en.wikipedia.org/w/index.php?title=File:Flag_of_Arizona.svg *License*: unknown *Contributors*: User:Madden

File:Flag of Minnesota.svg *Source*: http://en.wikipedia.org/w/index.php?title=File:Flag_of_Minnesota.svg *License*: unknown *Contributors*: Anime Addict AA, Denelson83, Dzordzm, Fry1989, Homo lupus, Interpretix, Jculbert, Juiced lemon, Serinde, Smooth O, SusanLesch, Svgalbertian, TownDown, Zscout370, 1 anonymous edits

File:Flag of Louisiana 1912.svg *Source*: http://en.wikipedia.org/w/index.php?title=File:Flag_of_Louisiana_1912.svg *License*: unknown *Contributors*: Cycn, Fry1989, Homo lupus, Hoshie, Infrogmation, Svgalbertian, Wst, Zscout370, 1 anonymous edits

File:Flag of Alabama.svg *Source*: http://en.wikipedia.org/w/index.php?title=File:Flag_of_Alabama.svg *License*: unknown *Contributors*: Steve Hall

File:Flag of Colorado.svg *Source*: http://en.wikipedia.org/w/index.php?title=File:Flag_of_Colorado.svg *License*: unknown *Contributors*: User:Fleminra

File:Flag of Kentucky.svg *Source*: http://en.wikipedia.org/w/index.php?title=File:Flag_of_Kentucky.svg *License*: unknown *Contributors*: Commonwealth of Kentucky

File:Flag of South Carolina.svg *Source*: http://en.wikipedia.org/w/index.php?title=File:Flag_of_South_Carolina.svg *License*: unknown *Contributors*: Abjiklam, Awg1010, Cwbm (commons), Cycn, Dbenbenn, Dzordzm, Fry1989, Homo lupus, Mattes, Nightstallion, Salah Almhamdi, Serinde, Smooth O, Zscout370, 3 anonymous edits

File:Flag of Oklahoma.svg *Source*: http://en.wikipedia.org/w/index.php?title=File:Flag_of_Oklahoma.svg *License*: unknown *Contributors*: Anime Addict AA, Denelson83, Dzordzm, Fry1989, Homo lupus, Pagrashtak, Rocket000, Svgalbertian, Trijnstel, Xenophon, Zscout370, 5 anonymous edits

File:Flag of Oregon.svg *Source*: http://en.wikipedia.org/w/index.php?title=File:Flag_of_Oregon.svg *License*: unknown *Contributors*: User:Denelson83

File:Flag of Connecticut.svg *Source*: http://en.wikipedia.org/w/index.php?title=File:Flag_of_Connecticut.svg *License*: unknown *Contributors*: Awg1010, CommonsDelinker, Denelson83, Dzordzm, Fry1989, Homo lupus, Mattes, Serinde, Smooth O, Svgalbertian, Zscout370, 5 anonymous edits

File:Flag of Iowa.svg *Source*: http://en.wikipedia.org/w/index.php?title=File:Flag_of_Iowa.svg *License*: unknown *Contributors*: Awg1010, Denelson83, Dzordzm, Erlenmeyer, Fry1989, Homo lupus, Hydrargyrum, Mattes, Mnmazur, Prenn, Serinde, Smooth O, Svgalbertian, 3 anonymous edits

File:Flag of Mississippi.svg *Source*: http://en.wikipedia.org/w/index.php?title=File:Flag_of_Mississippi.svg *License*: unknown *Contributors*: User:Pumbaa80

File:Flag of Kansas.svg *Source*: http://en.wikipedia.org/w/index.php?title=File:Flag_of_Kansas.svg *License*: unknown *Contributors*: AnonMoos, Denelson83, Dzordzm, Fry1989, Homo lupus, Mattes, Serinde, Smooth O, Svgalbertian, Theo10011, Zscout370, 2 anonymous edits

File:Flag of Arkansas.svg *Source*: http://en.wikipedia.org/w/index.php?title=File:Flag_of_Arkansas.svg *License*: unknown *Contributors*: Awg1010, Cycn, Dbenbenn, Dzordzm, Fry1989, Himasaram, Homo lupus, Nightstallion, Serinde, Smooth O, Vantey, Zscout370, 1 anonymous edits

File:Flag of Utah.svg *Source*: http://en.wikipedia.org/w/index.php?title=File:Flag_of_Utah.svg *License*: unknown *Contributors*: User:Svgalbertian, User:Zscout370

File:Flag of Nevada.svg *Source*: http://en.wikipedia.org/w/index.php?title=File:Flag_of_Nevada.svg *License*: unknown *Contributors*: Caleb Moore

File:Flag of New Mexico.svg *Source*: http://en.wikipedia.org/w/index.php?title=File:Flag_of_New_Mexico.svg *License*: unknown *Contributors*: AnonMoos, Bdk, Diltsgd, Dzordzm, Fry1989, Homo lupus, Mattes, Myself488, Nightstallion, Origamiemensch, Permjak, Rocket000, Serinde, Zscout370, 6 anonymous edits

File:Flag of West Virginia.svg *Source*: http://en.wikipedia.org/w/index.php?title=File:Flag_of_West_Virginia.svg *License*: unknown *Contributors*: User:Denelson83, User:Militaryace

File:Flag of Nebraska.svg *Source*: http://en.wikipedia.org/w/index.php?title=File:Flag_of_Nebraska.svg *License*: unknown *Contributors*: Ammodramus, Anime Addict AA, Denelson83, Dzordzm, Fry1989, Homo lupus, Ricordisamoa, Serinde, Svgalbertian, Zscout370, 2 anonymous edits

File:Flag of Idaho.svg *Source*: http://en.wikipedia.org/w/index.php?title=File:Flag_of_Idaho.svg *License*: unknown *Contributors*: Awg1010, Denelson83, Dzordzm, Fry1989, Homo lupus, Mmaxx, Orange Tuesday, Robbiegiles, Serinde, Smooth O, Svgalbertian, 2 anonymous edits

File:Flag of Maine.svg *Source*: http://en.wikipedia.org/w/index.php?title=File:Flag_of_Maine.svg *License*: unknown *Contributors*: A7x, Awg1010, Denelson83, Dzordzm, Fry1989, Himasaram, Homo lupus, Jmckean, Juiced lemon, Mattes, Serinde, Skeezix1000, Svgalbertian, 2 anonymous edits

File:Flag of New Hampshire.svg *Source*: http://en.wikipedia.org/w/index.php?title=File:Flag_of_New_Hampshire.svg *License*: unknown *Contributors*: Captain-tucker, Denelson83, Dzordzm, Fry1989, Homo lupus, Serinde, Smooth O, Svgalbertian, Theo10011, Zscout370, 1 anonymous edits

File:Flag of Hawaii.svg *Source*: http://en.wikipedia.org/w/index.php?title=File:Flag_of_Hawaii.svg *License*: unknown *Contributors*: Anime Addict AA, Awg1010, Cycn, Dbenbenn, Denelson83, Dzordzm, Editor at Large, F. F. Fjodor, Fry1989, Homo lupus, Hydrargyrum, Idh0854, KAVEBEAR, Kalathalan, Manuelt15, Mattes, Nagy, Nightstallion, Ricordisamoa, Serinde, Sinnamon, Svgalbertian, Telim tor, Vonvon, Yaddah, Zscout370, 19 anonymous edits

File:Flag of Rhode Island.svg *Source*: http://en.wikipedia.org/w/index.php?title=File:Flag_of_Rhode_Island.svg *License*: unknown *Contributors*: Awg1010, Cycn, Denelson83, Dzordzm, Fry1989, Homo lupus, Hydrargyrum, Juliancolton, Serinde, Smooth O, Svgalbertian, Zscout370, 1 anonymous edits

File:Flag of Montana.svg *Source*: http://en.wikipedia.org/w/index.php?title=File:Flag_of_Montana.svg *License*: unknown *Contributors*: Anime Addict AA, Denelson83, Dzordzm, Fry1989, Homo lupus, PumpkinSky, Serinde, Smooth O, Svgalbertian, Theo10011, Zscout370, 2 anonymous edits

File:Flag of Delaware.svg *Source*: http://en.wikipedia.org/w/index.php?title=File:Flag_of_Delaware.svg *License*: unknown *Contributors*: xrmap flag collection

File:Flag of South Dakota.svg *Source*: http://en.wikipedia.org/w/index.php?title=File:Flag_of_South_Dakota.svg *License*: unknown *Contributors*: Cycn, Denelson83, Dzordzm, Fry1989, Homo lupus, Serinde, Smooth O, Zscout370, 2 anonymous edits

File:Flag of North Dakota.svg *Source*: http://en.wikipedia.org/w/index.php?title=File:Flag_of_North_Dakota.svg *License*: unknown *Contributors*: Denelson83, Dzordzm, Fry1989, Homo lupus, Mattes, Myrtle T Turtle, Randbewohner, Serinde, Smooth O, Svgalbertian, Zscout370, 6 anonymous edits

File:Flag of Alaska.svg *Source*: http://en.wikipedia.org/w/index.php?title=File:Flag_of_Alaska.svg *License*: unknown *Contributors*: Anime Addict AA, AnonMoos, Cycn, Dbenbenn, DevinCook, Duck that quacks alot, Dzordzm, Fry1989, Homo lupus, Juiced lemon, Juliancolton, MGA73, Mattes, Nightstallion, R2D2Art2005, Resident Mario, Serinde, Smooth O, VIGNERON, Wester, Zscout370, 18 anonymous edits

File:Flag of Vermont.svg *Source*: http://en.wikipedia.org/w/index.php?title=File:Flag_of_Vermont.svg *License*: unknown *Contributors*: User:Denelson83

File:Flag of Washington, D.C..svg *Source*: http://en.wikipedia.org/w/index.php?title=File:Flag_of_Washington,_D.C..svg *License*: unknown *Contributors*: (of code)

File:Decrease2.svg *Source*: http://en.wikipedia.org/w/index.php?title=File:Decrease2.svg *License*: unknown *Contributors*: User:Sarang

File:Flag of Wyoming.svg *Source*: http://en.wikipedia.org/w/index.php?title=File:Flag_of_Wyoming.svg *License*: unknown *Contributors*: User:Dbenbenn

File:2000-census-percent-change.png *Source*: http://en.wikipedia.org/w/index.php?title=File:2000-census-percent-change.png *License*: unknown *Contributors*: Cholo Aleman, EurekaLott, Groupsixty, Ingolfson, Lensovet

File:2000-census-numeric-change.png *Source*: http://en.wikipedia.org/w/index.php?title=File:2000-census-numeric-change.png *License*: unknown *Contributors*: Cholo Aleman, EurekaLott, Groupsixty, Ingolfson, Lensovet

File:2000 census reapportionment.svg *Source*: http://en.wikipedia.org/w/index.php?title=File:2000_census_reapportionment.svg *License*: unknown *Contributors*: User:Kevthegreat55

File:Census2000Person2.png *Source*: http://en.wikipedia.org/w/index.php?title=File:Census2000Person2.png *License*: unknown *Contributors*: The Federal angency involved is the Census Bureau.

file:Census Bureau seal.svg *Source*: http://en.wikipedia.org/w/index.php?title=File:Census_Bureau_seal.svg *License*: unknown *Contributors*: United States Bureau of the CensusThis vector version by Mysid

file:U.S. Census Bureau logo post-2011.png *Source*: http://en.wikipedia.org/w/index.php?title=File:U.S._Census_Bureau_logo_post-2011.png *License*: unknown *Contributors*: U.S. federal government

Image:U.S. Census Bureau Regions.svg *Source*: http://en.wikipedia.org/w/index.php?title=File:U.S._Census_Bureau_Regions.svg *License*: unknown *Contributors*: U.S. Department of Commerce Economics and Statistics Administration U.S. CENSUS BUREAU

Image:New Regional Offices.png *Source*: http://en.wikipedia.org/w/index.php?title=File:New_Regional_Offices.png *License*: unknown *Contributors*: US Bureau of the Census

File:Census Bureau headquarters, Suitland, Maryland, 2007.jpg *Source*: http://en.wikipedia.org/w/index.php?title=File:Census_Bureau_headquarters,_Suitland,_Maryland,_2007.jpg *License*: unknown *Contributors*: United States Census Bureau

File:Census Regions and Division of the United States.svg *Source*: http://en.wikipedia.org/w/index.php?title=File:Census_Regions_and_Division_of_the_United_States.svg *License*: unknown *Contributors*: US Census Bureau

File:This is a card puncher, an integral part of the tabulation system used by the United States Census Bureau to compile... - NARA - 513295.tif *Source*: http://en.wikipedia.org/w/index.php?title=File:This_is_a_card_puncher,_an_integral_part_of_the_tabulation_system_used_by_the_United_States_Census_Bureau_to_compile..._-_NARA_-_513295.tif *License*: unknown *Contributors*: ArnoldReinhold, Infrogmation, Mattes, Umnik, Yann

File:Volkstelling 1925 Census.jpg *Source*: http://en.wikipedia.org/w/index.php?title=File:Volkstelling_1925_Census.jpg *License*: unknown *Contributors*: Aiko, Classical geographer, Frank C. Müller, JPS68, Vincent Steenberg, Vluebben

Image:Countries by population density.svg *Source*: http://en.wikipedia.org/w/index.php?title=File:Countries_by_population_density.svg *License*: unknown *Contributors*: Miguel Contreras, Guatemala

File:World_population_density_1994.png *Source*: http://en.wikipedia.org/w/index.php?title=File:World_population_density_1994.png *License*: unknown *Contributors*: User:QuartierLatin1968

Image:Population density with key.png *Source*: http://en.wikipedia.org/w/index.php?title=File:Population_density_with_key.png *License*: unknown *Contributors*: User:Keenan Pepper, User:SG

File:Aavikko.png *Source*: http://en.wikipedia.org/w/index.php?title=File:Aavikko.png *License*: unknown *Contributors*: User Vzb83 on fi.wikipedia

Image:Monaco by night.JPG *Source*: http://en.wikipedia.org/w/index.php?title=File:Monaco_by_night.JPG *License*: unknown *Contributors*: User:Ewa Dryjanska

File:TariatLandscape.jpg *Source*: http://en.wikipedia.org/w/index.php?title=File:TariatLandscape.jpg *License*: unknown *Contributors*: User:Doron

File:EU Pop2008 1024.PNG *Source*: http://en.wikipedia.org/w/index.php?title=File:EU_Pop2008_1024.PNG *License*: unknown *Contributors*: User:Vinny Burgoo

File:Flag of Singapore.svg *Source*: http://en.wikipedia.org/w/index.php?title=File:Flag_of_Singapore.svg *License*: unknown *Contributors*: Various

File:Flag of Hong Kong.svg *Source*: http://en.wikipedia.org/w/index.php?title=File:Flag_of_Hong_Kong.svg *License*: unknown *Contributors*: Designed by

File:Flag of Bahrain.svg *Source*: http://en.wikipedia.org/w/index.php?title=File:Flag_of_Bahrain.svg *License*: unknown *Contributors*: User:SKopp, User:Zscout370

File:Flag of Bangladesh.svg *Source*: http://en.wikipedia.org/w/index.php?title=File:Flag_of_Bangladesh.svg *License*: unknown *Contributors*: User:SKopp

File:Flag of the Republic of China.svg *Source*: http://en.wikipedia.org/w/index.php?title=File:Flag_of_the_Republic_of_China.svg *License*: unknown *Contributors*: 555, Abner1069, Bestalex, Bigmorr, Bjankuloski06en, Denelson83, Ed veg, Gzdavidwong, Herbythyme, Isletakee, Kakoui, Kallerna, Kibinsky, MAXXX-309, Mattes, Mizunoryu, Neq00, Nickpo, Nightstallion, Odder, Pymouss, R.O.C, Reisio, Reuvenk, Rkt2312, Rocket000, Runningfridgesrule, Samwingkit, Sasha Krotov, Shizhao, SiBr4, Tabasco, Theo10011, Vzb83, Wrightbus, ZooFari, Zscout370, 75 anonymous edits

File:Flag of Mauritius.svg *Source*: http://en.wikipedia.org/w/index.php?title=File:Flag_of_Mauritius.svg *License*: unknown *Contributors*: User:Zscout370

File:Flag of South Korea.svg *Source*: http://en.wikipedia.org/w/index.php?title=File:Flag_of_South_Korea.svg *License*: unknown *Contributors*: Various

File:Flag of Rwanda.svg *Source*: http://en.wikipedia.org/w/index.php?title=File:Flag_of_Rwanda.svg *License*: unknown *Contributors*: .

File:Flag of the Netherlands.svg *Source*: http://en.wikipedia.org/w/index.php?title=File:Flag_of_the_Netherlands.svg *License*: unknown *Contributors*: User:Zscout370

File:Flag of Lebanon.svg *Source*: http://en.wikipedia.org/w/index.php?title=File:Flag_of_Lebanon.svg *License*: unknown *Contributors*: Traced based on the CIA World Factbook with some modification done to the colours based on information at Vexilla mundi.

File:Flag of India.svg *Source*: http://en.wikipedia.org/w/index.php?title=File:Flag_of_India.svg *License*: unknown *Contributors*: Anomie, Mifter

File:Flag of Belgium (civil).svg *Source*: http://en.wikipedia.org/w/index.php?title=File:Flag_of_Belgium_(civil).svg *License*: unknown *Contributors*: Bean49, Cathy Richards, David Descamps, Dbenbenn, Denelson83, Evanc0912, Fry1989, Gabriel trzy, Howcome, IvanOS, Mimich, Ms2ger, Nightstallion, Oreo Priest, Ricordisamoa, Rocket000, Rodejong, SiBr4, Sir Iain, ThomasPusch, Warddr, Zscout370, 6 anonymous edits

File:Flag of Japan.svg *Source*: http://en.wikipedia.org/w/index.php?title=File:Flag_of_Japan.svg *License*: unknown *Contributors*: Anomie

File:Flag of Sri Lanka.svg *Source*: http://en.wikipedia.org/w/index.php?title=File:Flag_of_Sri_Lanka.svg *License*: unknown *Contributors*: Zscout370

File:800px-US Census 2000 race definitions Australia Sudan Afghan.PNG *Source*: http://en.wikipedia.org/w/index.php?title=File:800px-US_Census_2000_race_definitions_Australia_Sudan_Afghan.PNG *License*: unknown *Contributors*: Dark Tea, John of Reading

File:Percentage population living on less than $1.25 per day 2009.svg *Source*: http://en.wikipedia.org/w/index.php?title=File:Percentage_population_living_on_less_than_$1.25_per_day_2009.svg *License*: unknown *Contributors*: Tony0106

File:Percent poverty world map.png *Source*: http://en.wikipedia.org/w/index.php?title=File:Percent_poverty_world_map.png *License*: unknown *Contributors*: User:Sbw01f

Image:Time zone chicago.jpg *Source*: http://en.wikipedia.org/w/index.php?title=File:Time_zone_chicago.jpg *License*: unknown *Contributors*: User:JoeSmack

File:Standard time zones of the world.png *Source*: http://en.wikipedia.org/w/index.php?title=File:Standard_time_zones_of_the_world.png *License*: unknown *Contributors*: User:TimeZonesBoy

File:UTC hue4map X world Robinson.png *Source*: http://en.wikipedia.org/w/index.php?title=File:UTC_hue4map_X_world_Robinson.png *License*: unknown *Contributors*: User:Phoenix B 1of3

Image:Tzdiff-Europe-summer.png *Source*: http://en.wikipedia.org/w/index.php?title=File:Tzdiff-Europe-summer.png *License*: unknown *Contributors*: Jmabel, Naive cynic

Image:DaylightSaving-World-Subdivisions.png *Source*: http://en.wikipedia.org/w/index.php?title=File:DaylightSaving-World-Subdivisions.png *License*: unknown *Contributors*: Paul Eggert

File:Coventry Time Zone Clock.jpg *Source*: http://en.wikipedia.org/w/index.php?title=File:Coventry_Time_Zone_Clock.jpg *License*: unknown *Contributors*: User:Cmglee

MIX
Papier aus verantwortungsvollen Quellen
Paper from responsible sources
FSC® C105338

Printed by Books on Demand GmbH, Norderstedt / Germany